SCHAUM'S OUTLINE OF

THEORY AND PROBLEMS

OF

ELECTRIC POWER SYSTEMS

•

SYED A. NASAR, Ph.D.

Professor of Electrical Engineering
University of Kentucky

•

SCHAUM'S OUTLINE SERIES
McGRAW-HILL PUBLISHING COMPANY

New York St. Louis San Francisco Auckland Bogotá Caracas
Hamburg Lisbon London Madrid Mexico Montreal New Delhi
Oklahoma City Paris San Juan São Paulo Singapore Sydney
Tokyo Toronto

SYED A. NASAR received the Ph.D. degree in electrical engineering from the University of California at Berkeley. He is Professor of Electrical Engineering at the University of Kentucky, Lexington. He has been involved in teaching, research, and consulting in electrical machines for over 25 years. He is the author of two Schaum's Outlines, *Electric Machines and Electromechanics* and *Basic Electrical Engineering*. He is also the author or coauthor of 19 books and over 100 technical papers and is the editor of the monthly *Electric Machines and Power Systems*. Dr. Nasar received the Aurel Vlaicu award of the Romanian Academy of Science in 1978 for his contributions of linear machines. He is a Fellow IEEE and a Fellow IEE (London) and is a member of Eta Kappa Nu and Sigma Xi.

Schaum's Outline of Theory and Problems of
ELECTRIC POWER SYSTEMS

1 2 3 4 5 6 7 8 9 10 11 12 13 14 15 16 17 18 19 SHP SHP 8 9 2 1 0 9

ISBN 0-07-045917-7

Sponsoring Editor, David Beckwith
Production Supervisor, Denise Puryear
Editing Supervisor, Meg Tobin

Library of Congress Cataloging-in-Publication Data

Nasar, S. A.
    Schaum's outline of theory and problems of electric power systems
/ Syed A. Nasar.
        p.   cm. — (Schaum's outline series)
    Includes index.
    ISBN 0-07-045917-7
    1. Electric power systems.   I. Title.
TK1001.N38   1990
621.31—dc20   89-34889
                                                                       CIP

PREFACE

This book is written as a supplement to standard senior-level texts on electric power systems. However, certain topics, including growth rates, energy sources (Chapter 1), and underground cables (Chapter 5), that are not commonly found in most texts, are also discussed. Due to the nature of the book, detailed descriptive material and the derivations of most equations have been omitted. End results are given in analytic form and are illustrated with detailed numerical examples.

As prerequisites, the reader is expected to be familiar with ac circuits and electric machinery, especially transformers and synchronous machines.

The editorial help of Ed Millman is gratefully acknowledged.

Syed A. Nasar

CONTENTS

Chapter 1

Fundamentals of Electric Power Systems

The study of electric power systems is concerned with the generation, transmission, distribution, and utilization of electric power (Fig. 1-1). The first of these—the generation of electric power—involves the conversion of energy from a nonelectrical form (such as thermal, hydraulic, or solar energy) to electric energy. Thus, it is appropriate to begin this text with a discussion of energy.

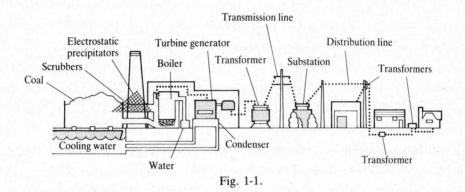

Fig. 1-1.

1.1 ENERGY AND POWER

Let a force F be applied to a mass so as to move the mass through a linear displacement l in the direction of F. Then the *work* U done by the force is defined as the product Fl; that is,

$$U = Fl \qquad (1.1)$$

If the displacement is not in the direction of F, then the work done is the product of the displacement and the component of the force along the displacement; that is,

$$U = Fl \cos \alpha \qquad (1.2)$$

where α is the angle that F makes with l. Work is measured in joules (J). From (1.1), one joule is the work done by a force of one newton in moving a body through a distance of one meter in the direction of the force: $1\,\text{J} = 1\,\text{N} \cdot \text{m}$.

The *energy* of a body is its capacity to do work. Energy has the same unit as work, although several other units are used for different forms of energy. For electric energy, the fundamental unit is the watt-second ($\text{W} \cdot \text{s}$), where

$$1\,\text{W} \cdot \text{s} = 1\,\text{J} \qquad (1.3)$$

More commonly, however, electric energy is measured in kilowatthours (kWh). From (1.3) we have

$$1\,\text{kWh} = 3.6 \times 10^6\,\text{J} \qquad (1.4)$$

The two most important forms of mechanical energy are kinetic energy and potential energy. A body possesses kinetic energy (KE) by virtue of its motion, such that an object of mass M (in kilograms), moving with a velocity u (in meters per second), has the kinetic energy

$$\text{KE} = \tfrac{1}{2}Mu^2 \quad \text{(in joules)} \qquad (1.5)$$

A body possesses potential energy (PE) by virtue of its position. Gravitational potential energy, for instance, results from an object's position in a gravitational field. A body of mass M (in

1

kilograms) at a height h (in meters) above the earth's surface has a gravitational PE given by

$$PE = Mgh \quad \text{(in joules)} \tag{1.6}$$

where g is the acceleration due to gravity, in meters per second per second.

Thermal energy is usually measured in calories (cal). By definition, one calorie is the amount of heat required to raise the temperature of one gram of water at 15°C through one Celsius degree. A more common unit is the kilocalorie (kcal). Experimentallly, it has been found that

$$1 \text{ cal} = 4.186 \text{ J} \tag{1.7}$$

Yet another unit of thermal energy is the British thermal unit (Btu), which is related to the joule and the calorie as follows:

$$1 \text{ Btu} = 1.055 \times 10^3 \text{ J} = 0.252 \times 10^3 \text{ cal} \tag{1.8}$$

Because the joule and the calorie are relatively small units, thermal energy and electric energy are generally expressed in terms of the British thermal unit and kilowatthour (or even megawatthour), respectively. A still larger unit of energy is the quad, which stands for "quadrillion British thermal units." The mutual relationships among these various units are

$$1 \text{ quad} = 10^{15} \text{ Btu} = 1.055 \times 10^{18} \text{ J} \tag{1.9}$$

(Some authors define 1 quad as 10^{18} Btu.)

Power is defined as the time rate at which work is done. Alternatively, power is the time rate of change of energy. Thus the instantaneous power p may be computed as

$$p = \frac{dU}{dt} = \frac{dw}{dt} \tag{1.10}$$

where U represents work and w represents energy. The SI unit of power is the watt (W); one watt is equivalent to one joule per second:

$$1 \text{ W} = 1 \text{ J/s} \tag{1.11}$$

Multiples of the watt commonly used in power engineering are the kilowatt and the megawatt. The power ratings (or outputs) of electric motors are expressed in horsepower (hp), where

$$1 \text{ hp} = 745.7 \text{ W} \tag{1.12}$$

1.2 GROWTH RATES

In planning to accommodate future electric energy needs, it is necessary that we have an estimate of the rate at which those needs will grow; Fig. 1-2 shows a typical energy-requirement projection for the United States.

Suppose a certain quantity M grows at a rate that is proportional to the amount of M that is present. Mathematically, we have

$$\frac{dM}{dt} = aM \tag{1.13}$$

where a is the constant of proportionality, known as the per-unit growth rate. The solution to (*1.13*) may be written as

$$M = M_0 e^{at} \tag{1.14}$$

where M_0 is the value of M at $t = 0$. At any two values of time, t_1 and t_2, the inverse ratio of the

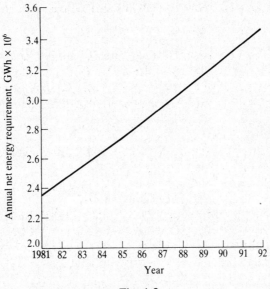

Fig. 1-2.

corresponding quantities M_1 and M_2 is

$$\frac{M_2}{M_1} = e^{a(t_2 - t_1)} \tag{1.15}$$

From (1.15) we may obtain the *doubling time* t_d such that $M_2 = 2M_1$ and $t_2 - t_1 = t_d$. It is

$$t_d = \frac{\ln 2}{a} = \frac{0.693}{a} \tag{1.16}$$

Power system planners also need to know how much power will be demanded. The peak power demand for the United States over several years is shown by the solid curve in Fig. 1-3. We can

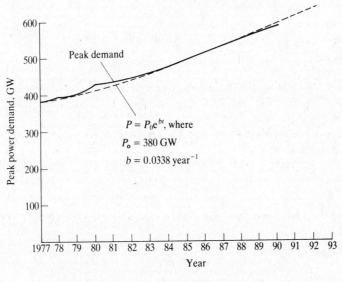

Fig. 1-3.

approximate this curve with the curve whose equation is

$$P = P_0 e^{bt} \qquad (1.17)$$

(dashed in Fig. 1-3), where P_0 is the peak power at time $t = 0$, and b is the per-unit growth rate for peak power. The area under this curve over a given period is a measure of the energy Q consumed during that period.

From (1.16) and (1.17) it follows that if the per-unit growth rate has not changed, then the energy consumed in one doubling period equals the energy consumed for the entire time prior to that doubling period. In particular, we obtain

$$Q_1 = Q_2 = \frac{P_0}{b} e^{bt_1} \qquad (1.18)$$

where Q_1 is the energy consumed up to a certain time t_1, Q_2 is the energy consumed during the doubling time t_d, and b is the per-unit power growth rate.

1.3 MAJOR ENERGY SOURCES

Fossil fuels—coal, petroleum, and natural gas—are major sources of energy for the generation of electric power. Another major source of energy on the earth is solar radiation, which may be obtained either directly as intercepted solar radiation or indirectly as wind and hydropower. Other significant forms of energy are tidal energy, geothermal energy, and nuclear energy.

Turbine-type *wind-energy* generators transform the kinetic energy of the wind into rotary-shaft motion and, in turn, to electrical energy. The power that can be extracted from wind is given approximately by

$$P = 2.46 \times 10^{-3} D^2 u^2 \quad \text{(in watts)} \qquad (1.19)$$

where D is the blade diameter in feet, and u is the wind velocity in miles per hour.

In *hydropower* conversion, the potential energy of a mass of water at a hydraulic head is converted into the kinetic energy of a hydraulic turbine that drives an electric generator. By (1.6), the potential energy of 1000 kg of water at a head of 100 m is 9.8×10^5 J. Alternatively, a flow rate of $1\,\text{m}^3/\text{s}$ with a head of 100 m provides $9.8 \times 10^3 \times \text{head} = 9.8 \times 10^3 \times 100 = 9.8 \times 10^5$ W of hydraulic power.

Tidal energy is obtained by closing off a bay with a dam, allowing it to fill during periods of high tide, and recovering the energy as it empties during periods of low tide. For a maximum tidal head H (in meters), the average tidal power obtained per unit area of tidal bay is given approximately by

$$P_{av} = 0.219H^2 \quad \text{(in megawatts per square kilometer)} \qquad (1.20)$$

Solved Problems

1.1 The net energy requirement for the United States in 1986 was approximately 2.82×10^6 GWh. What is the equivalent of this energy in British thermal units?

Since

$$1\,\text{GWh} = 10^9\,\text{Wh} = 10^6\,\text{kWh}$$

we have

$$2.82 \times 10^6\,\text{GWh} = 2.82 \times 10^{12}\,\text{kWh}$$

Then, from (1.4),

$$2.82 \times 10^{12}\,\text{kWh} = 3.6 \times 10^6 \times 2.82 \times 10^{12}\,\text{J} = 10.152 \times 10^{18}\,\text{J}$$

From (1.8) and (1.9) we finally obtain

$$10.152 \times 10^{18} \, J = \frac{10.152}{1.055} \times 10^{15} \, Btu = 9.623 \times 10^{15} \, Btu$$

$$= 9.623 \, quad$$

1.2   Coal has an average energy content of $940 \, W \cdot years/ton$, and natural gas has an energy content of $0.036 \, W \cdot year/ft^3$. If 80 percent of the net energy requirement of Problem 1.1 were to be met with coal and 20 percent with gas, what amounts of coal and gas would be required?

From Problem 1.1,

$$2.82 \times 10^6 \, GWh = \frac{2.82 \times 10^{15}}{365 \times 24} \, W \cdot years = 3.22 \times 10^{11} \, W \cdot years$$

Hence, we have

$$Energy \ to \ be \ supplied \ by \ coal = 0.8 \times 3.22 \times 10^{11} = 2.576 \times 10^{11} \, W \cdot years$$
$$Energy \ to \ be \ supplied \ by \ gas = 0.2 \times 3.22 \times 10^{11} = 6.44 \times 10^{10} \, W \cdot years$$

which lead to

$$Amount \ of \ coal \ required = \frac{2.567 \times 10^{11}}{940} = 2.74 \times 10^8 \, tons$$

$$Amount \ of \ gas \ required = \frac{6.44 \times 10^{10}}{0.036} = 1.79 \times 10^{12} \, ft^3$$

1.3   A certain amount of fuel can be converted into 3×10^{-3} quads of energy in a power station. If the average load on the station over a 24-h period is 50 MW, determine how long (in days) the fuel will last. Assume a 20 percent overall efficiency for the power station.

From (1.9) and (1.11), the energy available from the fuel is

$$3 \times 10^{-3} \, quad = 3 \times 10^{-3} \times 1.055 \times 10^{18} \, W \cdot s$$

$$= \frac{3 \times 10^{-3} \times 1.055 \times 10^{18}}{60 \times 60 \times 10^6} \, MWh = 8.79 \times 10^5 \, MWh$$

In 24 h, the station produces $50 \times 24 = 1200 \, MWh$ of energy. At 20 percent efficiency, this requires a daily energy input (from the fuel) of $1200/0.2 = 6000 \, MWh$. Hence, the fuel will be consumed in $8.79 \times 10^5/6000 = 146.5$ days.

1.4   In 1981, the U. S. consumption of energy (in quads) from various sources was as follows: coal, 16.1; oil 32.1; natural gas, 20.2; hydro, 2.9; and nuclear, 2.9. Calculate in gigawatthours the total electric energy that could be produced from these sources, assuming an average power-plant conversion efficiency of 0.1.

The total amount of energy consumed in 1981 was

$$16.1 + 32.1 + 20.2 + 2.9 + 2.9 = 74.2 \, quad$$

$$= \frac{74.2 \times 1.055 \times 10^{18}}{3.6 \times 10^6} \, kWh$$

$$= 21.75 \times 10^6 \, GWh$$

At an efficiency of 0.1, this produced $2.175 \times 10^6 \, GWh$ of electric energy.

1.5   The average heat content of natural gas is $1.05 \, Btu/ft^3$, and that of bituminous coal is

14,000 Btu/lb. Using the data of Problem 1.4, determine the amounts of natural gas and coal consumed in the United States in 1981.

Since 20.2 quads were derived from natural gas, we have

$$20.2 \text{ quads} = 20.2 \times 10^{15} \text{ Btu}$$

and

$$\frac{20.2}{1.05} \times 10^{15} = 19.238 \times 10^{15} \text{ ft}^3$$

Similarly, for the coal we have

$$16.1 \text{ quads} = 16.1 \times 10^{15} \text{ Btu}$$

and

$$\frac{16.1 \times 10^{15}}{14,000} = 1.15 \times 10^{12} \text{ lb} \quad \text{or} \quad 5.75 \times 10^8 \text{ tons}$$

1.6　　Suppose that the consumption of energy in a certain country has a growth rate of 4 percent per year. In how many years will the energy consumption be tripled?

From (*1.15*) with $Q_2/Q_1 = 3$,

$$3 = e^{0.04t} \quad \text{or} \quad \ln 3 = 0.04t$$

and

$$t = \frac{\ln 3}{0.04} = 27.47 \text{ years}$$

1.7　　In a certain country the energy consumption is expected to double in 10 years. Calculate the growth rate.

From (*1.17*),

$$a = \frac{0.693}{10} = 6.93 \text{ percent}$$

1.8　　Derive (*1.18*).

Evaluating the energy Q_1 consumed up to time t_1 and the energy Q_2 consumed during the doubling time $t_d = t_2 - t_1$, we obtain, from (*1.17*),

$$Q_1 = \int_{-\infty}^{t_1} P_0 e^{bt} \, dt = \frac{P_0}{b} e^{bt_1}$$

and

$$Q_2 = \int_{t_1}^{t_2} P_0 e^{bt} \, dt = \frac{P_0}{b} (e^{bt_d} - 1) e^{bt_1}$$

From (*1.16*), $t_d = (\ln 2)/b$, so Q_2 becomes

$$Q_2 = \frac{P_0}{b} (2 - 1) e^{bt_1} = \frac{P_0}{b} e^{bt_1} = Q_1$$

1.9　　Sketch a curve showing doubling time (in years) as a function of growth rate (in percent per year). From the graph, obtain the doubling time for a growth rate of 5 percent.

The graph, shown in Fig. 1-4, is plotted by use of (*1.16*). It shows that $t_d \approx 14$ years. Applying (*1.16*) directly yields $t_d = 0.693/0.05 = 13.86$ years.

1.10　　The present estimate of solid coal reserves in the United States is about 1.5×10^9 tons, with an energy content of 940 W · years/ton. If the power consumption growth rate is 3.38 percent

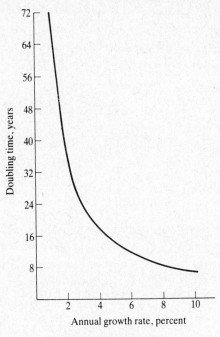

Fig. 1-4.

per year, approximately how long will the coal reserves last? Assume that all the energy will be supplied by coal and that the present peak power demand is 425 GW.

Let T be the time at which total consumption will equal the reserves and Q_T the total energy reserve. Then with $t = 0$ at the present time, we have

$$Q_T = \int_0^T P_0 e^{bt}\, dt = \frac{P_0}{b}(e^{bT} - 1)$$

which we may write as

$$e^{bT} = \frac{bQ_T}{P_0} + 1$$

so that

$$T = \frac{1}{b}\ln\left(\frac{bQ_T}{P_0} + 1\right) \tag{1}$$

From the given numerical values we have

$$Q_T = 1.5 \times 10^9 \times 940 = 1.41 \times 10^{12}\,\text{W} \cdot \text{years}$$

and

$$T = \frac{1}{0.0338}\ln\left(\frac{0.0338 \times 1.41 \times 10^{12}}{425 \times 10^9} + 1\right) = 3.144\ \text{years}$$

1.11   Express (*1.19*) in SI units.

Since $3.28\,\text{ft} = 1\,\text{m}$ and $1\,\text{m/s} = 2.237\,\text{mi/h}$, (*1.19*) becomes

$$P = 2.46 \times 10^{-3}(3.28D)^2(2.237u)^3$$
$$= 0.29626D^2U^3 \quad \text{(in watts)}$$

where D is in meters and U is in meters per second.

1.12   A small wind generator is designed to generate 50 kW of power at a wind velocity of 25 mi/h.

What is the approximate blade diameter?

From (1.19),

$$50 \times 10^3 = 2.46 \times 10^{-3}D^2(25)^3$$

so

$$D = \sqrt{\frac{50 \times 10^3}{2.46 \times 10^{-3} \times (25)^3}} = 36 \, \text{ft}$$

1.13 The wind velocity operating the generator of Problem 1.12 actually varies between 20 and 50 km/h. Determine the range of available power.

At 20 km/h (or 12.4 mi/h),

$$P_{20} = 2.46 \times 10^{-3}(36)^2(12.4)^3 = 6.08 \, \text{kW}$$

At 50 km/h (or 31 mi/h),

$$P_{50} = 2.46 \times 10^{-3}(36)^2(31)^3 = 94.98 \, \text{kW}$$

Hence, approximately, $6 < P < 95 \, \text{kW}$.

1.14 One million cubic meters of water is stored in a reservoir feeding a water turbine. If the center of mass of the water is 50 m above the turbine and losses are negligible, how much energy (in megawatthours) will that volume of water produce? The density of water is 993 kg/m³.

The weight of the water is 993×10^6 kg. By (1.6), its potential energy is

$$PE = 993 \times 10^6 \times 9.81 \times 50 \, \text{W} \cdot \text{s}$$

$$= \frac{993 \times 10^6 \times 9.81 \times 50}{3600 \times 10^6} = 135.3 \, \text{MWh}$$

1.15 Coal reserves in the eastern United States are estimated to contain 2250 quads of energy. If the energy content of this coal is 11,500 Btu/lb, determine the approximate weight of the coal reserve.

$$\text{Approximate weight} = \frac{2250 \times 10^{15}}{11,500 \times 2000} = 9.78 \times 10^{10} \, \text{tons}$$

1.16 A power plant consumes 3600 tons of coal per day. If the coal has an average energy content of 10,000 Btu/lb, what is the plant's power output? Assume an overall efficiency of 15 percent.

The power available from the coal is

$$\frac{3600 \times 2000 \times 10,000}{24} = 3 \times 10^9 \, \text{Btu/h}$$

In megawatts, this is

$$\frac{3 \times 10^9 \times 1.055 \times 10^3}{60 \times 60 \times 10^6} = 879 \, \text{MW}$$

At 15 percent efficiency,

$$\text{Power output} = 0.15 \times 879 = 132 \, \text{MW}$$

1.17 Present natural gas reserves in the United States are estimated to contain 452 quads of energy. The present peak electric power demand is 450 GW. If the power consumption growth rate is 6.5 percent per year, and 22 percent of the total energy consumption is to be supplied by natural gas, approximately how long will these natural gas reserves last?

In the nomenclature of Problem 1.10, we have

$$Q_T = \frac{452 \times 1.055 \times 10^{18}}{365 \times 24 \times 3600} = 1.512 \times 10^{13} \text{ W} \cdot \text{years}$$

$$b = 0.065$$

$$P_0 = 450 \times 0.22 \times 10^9 = 9.9 \times 10^{10} \text{ W}$$

and

$$\frac{bQ_T}{P_0} = \frac{0.065 \times 1.512 \times 10^{13}}{9.9 \times 10^{10}} = 9.928$$

Then

$$T = \frac{1}{b} \ln \left(\frac{bQ_T}{P_0} + 1 \right) = \frac{1}{0.065} \ln (9.928 + 1) = 36.8 \text{ years}$$

1.18 Estimate the average power output of a wind turbine having a blade diameter of 35 ft if the wind velocity ranges from 10 to 30 mi/h.

From (*1.19*),

$$P_{min} = 2.46 \times 10^{-3} \times 35^2 \times 10^3 = 2.46 \times 35^2 \text{ W}$$

and

$$P_{max} = 2.46 \times 10^{-3} \times 35^2 \times 30^3 = 2.46 \times 35^2 \times 27 \text{ W}$$

Then

$$P_{av} = 1.23 \times 35^2 \times 28 = 42.2 \text{ kW}$$

1.19 The maximum tidal head available for a proposed tidal-power station is 6 m. What must be the area of the tidal bay to generate an average of 1000 MW of power?

From (*1.20*), we have

$$1000 = 0.219 \times 36 \times \text{area}$$

so

$$\text{Area} \doteq \frac{1000}{0.219 \times 36} = 126.8 \text{ m}^2$$

Supplementary Problems

1.20 A certain amount of fuel contains 15×10^{10} Btu of energy. What is the corresponding energy in kilocalories?

Ans. 3.78×10^{10} kcal

1.21 The fuel of Problem 1.20 is converted into electric energy in a power station having a 12 percent overall efficiency. The average demand on the station over a 24-h period is 5 MW. In how many days will the fuel be totally consumed?

Ans. 44 days.

1.22 A certain amount of fuel can produce 10 quads of energy. In how many days will the fuel be totally consumed if it is used to satisfy a demand of 10^{13} Btu/day at a power plant with an overall efficiency of 20 percent?

Ans. 200 days

1.23 Calculate the total energy (in kilocalories) available from the fuel of Problem 1.22.

Ans. 2.52×10^{15} kcal

1.24 A 90 percent efficient electric motor operating an elevator lifts a 10-ton load through a height of 60 ft. Calculate the energy required by the motor to do so.

Ans. 1.81 MJ

1.25 The load of Problem 1.22 is to be lifted through the entire height of 60 ft within 40 s. Determine the minimum horsepower rating of the motor.

Ans. 55 hp (approximately)

1.26 Calculate the energy required for a geared dc motor to lift a 1-ton load through 50 ft in 10 s. The motor/gear overall efficiency is 0.51.

Ans. 0.083 kWh

1.27 An electric hoist makes 10 round trips per hour. For each trip, a load of 6 tons is raised in a hoist cage to a height of 200 ft in $1\frac{1}{2}$ min, and then the cage returns empty in $1\frac{1}{4}$ min. The cage weighs 0.5 ton and has a balance weight of 3 tons. The efficiency of the hoist is 80 percent, and that of the driving motor is 88 percent. Calculate the electric energy required per round trip.

Ans. 1.44 kWh

1.28 A belt-driven generator supplies 875 kW at 95 percent efficiency. If the loss in the drive belt is 2.5 percent, calculate the horsepower of the engine needed to drive the generator.

Ans. 1267 hp

1.29 In a power station, 4×10^4 GWh of energy is to be produced in 1 year, half from coal and half from natural gas. The energy content of coal is 900 W · years/ton, and that of natural gas is 0.03 W · year/ft³. How much coal and how much natural gas will be required?

Ans. 2.537×10^6 tons; 76.1×10^9 ft³

1.30 Rework Problem 1.24 assuming all the energy is to be supplied by (*a*) coal and (*b*) natural gas.

Ans. (*a*) 5.0735×10^6 tons; (*b*) 152.2×10^9 ft³

1.31 During a 1-year period, a certain power system consumed energy (in quads) from various sources as follows: coal, 6; oil, 2; gas, 1; and hydro, 0.5. If the overall efficiency of the system is 0.12, how much electric energy (in gigawatthours) could be produced by the system from these sources?

Ans. 3.2×10^6 GWh

1.32 In a certain region the growth rate of energy consumption is 6 percent. In how many years will the energy consumption be quadrupled?

Ans. 23.1 years

1.33 Natural gas reserves in a certain country are estimated at 100×10^9 ft³, with an energy content of 0.025 W · year/ft³. If the present peak power demand is 0.5 GW, the power demand growth rate is 5 percent, and all the energy is to be supplied by natural gas, approximately how long will the reserve last?

Ans. 4.46 years

1.34 Calculate the velocity with which a 200-kg mass must move so that its kinetic energy equals the energy dissipated in a 0.2-Ω resistor through which a 100-A current flows for 2 h.

Ans. 379.5 m/s

1.35   A wind generator with an efficiency of 0.85 has a blade diameter of 20 m. If the wind velocity is 30 km/h, how much power is obtainable from the generator?

Ans.   58.3 kW

1.36   Hydroelectric power is generated at a dam that produces a head of 180 ft and a reservoir containing 3×10^6 gal of water. How much energy can be generated from this reservoir by a turbine-generator system whose overall efficiency is 20 percent?

Ans.   1225.3 MJ

1.37   The reservoir of a hydroelectric generating station measures 217.8 ft by 200 ft at the surface. Its head decreases by 1 ft while the station generates 100 hp at 70 percent efficiency. Find the original head in feet.

Ans.   104 ft

1.38   A hydroelectric generating station is supplied from a reservoir of capacity 2×10^8 ft^3 at a head of 500 ft. What is the total available electric energy in kilowatthours if the hydraulic efficiency is 0.8 and the electrical efficiency is 0.9?

Ans.   1690 MWh

1.39   In a certain country the equivalent fuel reserve for power generation is 3×10^6 MW · years. The present peak power demand is 200 GW, and the expected power consumption growth rate is 2.1 percent. How long will the fuel reserve last?

Ans.   13 years

Chapter 2

Power System Representation

The basic components of a power system are generators, transformers, transmission lines, and loads. The interconnections among these components in the power system may be shown in a so-called one-line diagram. For analysis, the equivalent circuits of the components are shown in a reactance diagram or an impedance diagram.

2.1 ONE-LINE DIAGRAMS

Figure 2-1 shows the symbols used to represent the typical components of a power system. Figure 2-2 is a *one-line diagram* for a power system consisting of two generating stations connected by a transmission line; note the use of the symbols of Fig. 2-1. The advantage of such a one-line representation is its simplicity: One phase represents all three phases of the balanced system; the equivalent circuits of the components are replaced by their standard symbols; and the completion of the circuit through the neutral is omitted.

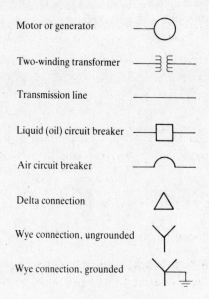

Motor or generator

Two-winding transformer

Transmission line

Liquid (oil) circuit breaker

Air circuit breaker

Delta connection

Wye connection, ungrounded

Wye connection, grounded

Fig. 2-1.

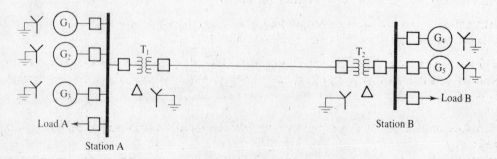

Fig. 2-2.

12

2.2 IMPEDANCE AND REACTANCE DIAGRAMS

The one-line diagram may serve as the basis for a circuit representation that includes the equivalent circuits of the components of the power system. Such a representation is called an *impedance diagram*, or a *reactance diagram* if resistances are neglected. The impedance and reactance diagrams corresponding to Fig. 2-2 are shown in Fig. 2-3(*a*) and (*b*), respectively. Note that only a single phase is shown.

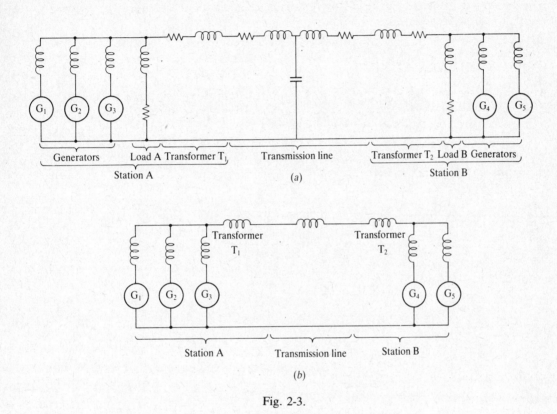

Fig. 2-3.

The following assumptions have been incorporated into Fig. 2-3(*a*):

1. A generator can be represented by a voltage source in series with an inductive reactance. The internal resistance of the generator is negligible compared to the reactance.

2. The loads are inductive.

3. The transformer core is ideal, and the transformer may be represented by a reactance.

4. The transmission line is a medium-length line and can be denoted by a T circuit. An alternative representation, such as a π circuit, is equally applicable.

5. The delta-wye-connected transformer T_1 may be replaced by an equivalent wye-wye-connected transformer (via a delta-to-wye transformation) so that the impedance diagram may be drawn on a per-phase basis.

(The exact nature and values of the impedances or reactances are determined by methods discussed in later chapters.)

The reactance diagram, Fig. 2-3(*b*), is drawn by neglecting all resistances, the static loads, and the capacitance of the transmission line.

2.3 PER-UNIT REPRESENTATION

Computations for a power system having two or more voltage levels become very cumbersome when it is necessary to convert currents to a different voltage level wherever they flow through a transformer (the change in current being inversely proportional to the transformer turns ratio). In an alternative and simpler system, a set of base values, or *base quantities,* is assumed for each voltage class, and each parameter is expressed as a decimal fraction of its respective base. For instance, suppose a base voltage of 345 kV has been chosen, and under certain operating conditions the actual system voltage is 334 kV; then the ratio of actual to base voltage is 0.97. The actual voltage may then be expressed as 0.97 *per-unit.* In an equally common practice, per-unit quantities are multiplied by 100 to obtain percent quantities; our example voltage would then be expressed as 97 *percent.*

Per-unit and percent quantities and their bases exhibit the same relationships and obey the same laws (such as Ohm's law and Kirchhoff's laws) as do quantities in other systems of units.

A minimum of four base quantities is required to completely define a per-unit system; these are voltage, current, power, and impedance (or admittance). If two of them are set arbitrarily, then the other two become fixed. The following relationships hold on a per-phase basis:

$$\text{Base current} = \frac{\text{base voltamperes}}{\text{base voltage}} \quad \text{(in amperes)} \tag{2.1}$$

$$\text{Base impedance} = \frac{\text{base voltage}}{\text{base current}} \quad \text{(in ohms)} \tag{2.2}$$

$$\text{Per-unit voltage} = \frac{\text{actual voltage}}{\text{base voltage}} \quad \text{(per unit, or pu)} \tag{2.3}$$

$$\text{Per-unit current} = \frac{\text{actual current}}{\text{base current}} \quad \text{(per unit, or pu)} \tag{2.4}$$

$$\text{Per-unit impedance} = \frac{\text{actual impedance}}{\text{base impedance}} \quad \text{(per unit, or pu)} \tag{2.5}$$

In a three-phase system, the base kVA may be chosen as the three-phase kVA, and the base voltage as the line-to-line voltage; or, the base values may be taken as the phase quantities. In either case, the per-unit three-phase kVA and voltage on the three-phase kVA base and the per-unit per-phase kVA and voltage on the kVA-per-phase base remain the same.

2.4 CHANGE OF BASE

The per-unit (pu) impedance of a generator or transformer, as supplied by the manufacturer, is generally based on the rating of the generator or transformer itself. However, such a per-unit impedance can be referred to a new voltampere base with the equation

$$(\text{Per-unit impedance})_{\text{new base}} = \frac{(\text{VA})_{\text{new base}}(\text{kV})^2_{\text{old base}}}{(\text{VA})_{\text{old base}}(\text{kV})^2_{\text{new base}}}(\text{per-unit impedance})_{\text{old base}} \tag{2.6}$$

If the old base voltage and new base voltage are the same, then (2.6) simplifies to

$$(\text{Per-unit impedance})_{\text{new base}} = \frac{(\text{VA})_{\text{new base}}}{(\text{VA})_{\text{old base}}}(\text{per-unit impedance})_{\text{old base}} \tag{2.7}$$

The impedances of transmission lines are expressed in ohms, but can be easily converted to pu values on a given voltampere base using (2.1) to (2.5).

2.5 SUMMARY OF THREE-PHASE CIRCUIT RELATIONSHIPS

A three-phase circuit may be connected either in wye or in delta. In a balanced three-phase circuit the phase and the line values of the current, power, and voltage are related as follows (the subscripts p and l refer to phase and line values, respectively):

Wye connection:

$$I_p = I_l$$
$$V_p = V_l/\sqrt{3}$$
$$P = \sqrt{3}V_lI_l\cos\theta_p$$

Delta connection:

$$I_p = I_l/\sqrt{3}$$
$$V_p = V_l$$
$$P = \sqrt{3}V_lI_l\cos\theta_p$$

The delta and wye impedances are related by

$$Z_{\text{wye}} = \tfrac{1}{3}Z_{\text{delta}}$$

For both types of connections, the apparent and reactive powers are, respectively,

$$\text{VA} = \sqrt{3}V_lI_l$$

and

$$Q = \sqrt{3}V_lI_l\sin\theta_p$$

From the above, it is clear that the phrase angle may be obtained as

$$\tan\theta_p = \frac{Q}{P}$$

Solved Problems

2.1 The base impedance and base voltage for a given power system are $10\,\Omega$ and $400\,\text{V}$, respectively. Calculate the base kVA and the base current.

From Ohm's law,

$$\text{Base current} = \frac{400}{10} = 40\,\text{A}$$

$$\text{Base kVA} = \frac{40 \times 400}{1000} = 16\,\text{kVA}$$

2.2 The base current and base voltage of a 345-kV system are chosen to be $3000\,\text{A}$ and $300\,\text{kV}$, respectively. Determine the per-unit voltage and the base impedance for the system.

From (*2.2*),

$$\text{Base impedance} = \frac{300 \times 10^3}{3000} = 100\,\Omega$$

From (*2.3*),

$$\text{Per-unit voltage} = \frac{345}{300} = 1.15\,\text{pu}$$

2.3   If the rating of the system of Problem 2.2 is 1380 MVA, calculate the per-unit current referred to the base of Problem 2.2.

We need the actual current in the system:

$$\text{Actual current} = \frac{1380 \times 10^6}{345 \times 10^3} = 4000 \text{ A}$$

Then, from (2.4),

$$\text{Per-unit current} = \frac{4000}{3000} = 1.33 \text{ pu}$$

2.4   Express a 100-Ω impedance, a 60-A current, and a 220-V voltage as per-unit quantities referred to the base values of Problem 2.1.

From (2.5),

$$\text{Per-unit impedance} = \frac{100}{10} = 10 \text{ pu}$$

From (2.4),

$$\text{Per-unit current} = \frac{60}{40} = 1.5 \text{ pu}$$

From (2.3),

$$\text{Per-unit voltage} = \frac{220}{400} = 0.55 \text{ pu}$$

2.5   A single-phase, 10-kVA, 200-V generator has an internal impedance Z_g of 2 Ω. Using the ratings of the generator as base values, determine the generated per-unit voltage that is required to produce full-load current under short-circuit conditions.

In per-unit terms, we have

$$\text{Base voltage} = 200 \text{ V} = 1 \text{ pu}$$
$$\text{Base kVA} = 10 \text{ kVA} = 1 \text{ pu}$$

Then, by (2.1),

$$\text{Base current} = \frac{10{,}000}{200} = 50 \text{ A} = 1 \text{ pu}$$

The generated voltage required to produce the rated current under short circuit is $IZ_g = 50 \times 2 = 100$ V; or, in per-unit terms, $100/200 = 0.5$ pu.

2.6   Let a 5-kVA, 400/200-V transformer be approximately represented by a 2-Ω reactance referred to the low-voltage side. Considering the rated values as base quantities, express the transformer reactance as a per-unit quantity.

We have

$$\text{Base voltamperes} = 5000 \text{ VA} \qquad \text{and} \qquad \text{base voltage} = 200 \text{ V}$$

so that, by (2.1) and (2.2),

$$\text{Base current} = \frac{5000}{200} = 25 \text{ A}$$

$$\text{Base impedance} = \frac{200}{25} = 8 \text{ }\Omega$$

Then the per-unit reactance referred to the low-voltage side is

$$\text{Per-unit reactance} = \frac{2}{8} = 0.25 \text{ pu}$$

2.7 Repeat Problem 2.6, expressing all quantities in terms of the high-voltage side.

Here we have

$$\text{Base voltamperes} = 5000 \text{ VA} \qquad \text{and} \qquad \text{base voltage} = 400 \text{ V}$$

Hence,

$$\text{Base current} = \frac{5000}{400} = 12.5 \text{ A}$$

and

$$\text{Base impedance} = \frac{400}{12.5} = 32 \,\Omega$$

The transformer reactance referred to the high-voltage side is

$$\text{High-side reactance} = 2\left(\frac{400}{200}\right)^2 = 8 \,\Omega$$

and the per-unit high-side reactance is $8/32 = 0.25$ pu.

2.8 Express the per-unit impedance Z_{pu} and per-unit admittance Y_{pu} of a power system in terms of the base voltage V_{base} and the base voltamperes $(\text{VA})_{\text{base}}$.

From (2.2), the base impedance is

$$Z_{\text{base}} = \frac{\text{base voltage}}{\text{base current}} = \frac{V_{\text{base}}}{(\text{VA})_{\text{base}}/V_{\text{base}}} = \frac{V_{\text{base}}^2}{(\text{VA})_{\text{base}}}$$

Then, from (2.5), the per-unit impedance is

$$Z_{\text{pu}} = \frac{\text{actual impedance}}{\text{base impedance}} = \frac{Z}{Z_{\text{base}}} = \frac{Z(\text{VA})_{\text{base}}}{V_{\text{base}}^2} \qquad \text{pu}$$

The per-unit admittance is

$$Y_{\text{pu}} = \frac{1}{Z_{\text{pu}}} = \frac{Y V_{\text{base}}^2}{(\text{VA})_{\text{base}}} \qquad \text{pu}$$

2.9 A 345-kV transmission line has a series impedance of $(4 + j60)\,\Omega$ and a shunt admittance of $j2 \times 10^{-3}$ S. Using 100 MVA and the line voltage as base values, calculate the per-unit impedance and per-unit admittance of the line.

From the results of Problem 2.8, we have

$$Z_{\text{pu}} = (4 + j60)\frac{100 \times 10^6}{(345 \times 10^3)^2} = (3.36 + j50.4) \times 10^{-3} \text{ pu}$$

$$Y_{\text{pu}} = (j2 \times 10^{-3})\frac{(345 \times 10^3)^2}{100 \times 10^6} = j2.38 \text{ pu}$$

2.10 A three-phase, wye-connected system is rated at 50 MVA and 120 kV. Express 40,000 kVA of three-phase apparent power as a per-unit value referred to (*a*) the three-phase system kVA as base and (*b*) the per-phase system kVA as base.

(*a*) For the three-phase base,

$$\text{Base kVA} = 50,000 \text{ kVA} = 1 \text{ pu}$$

and
$$\text{Base kV} = 120 \text{ kV (line to line)} = 1 \text{ pu}$$

so
$$\text{Per-unit kVA} = \frac{40,000}{50,000} = 0.8 \text{ pu}$$

(b) For the per-phase base,

$$\text{Base kVA} = \tfrac{1}{3} \times 50,000 = 16,667 = 1 \text{ pu}$$

and
$$\text{Base kV} = \frac{120}{\sqrt{3}} = 69.28 \text{ kV} = 1 \text{ pu}$$

so
$$\text{Per-unit kVA} = \frac{1}{3} \times \frac{40,000}{16,667} = 0.8 \text{ pu}$$

2.11 A three-phase, wye-connected, 6.25-kVA, 220-V synchronous generator has a reactance of 8.4 Ω per phase. Using the rated kVA and voltage as base values, determine the per-unit reactance. Then refer this per-unit value to a 230-V, 7.5-kVA base.

For the first base, we have

$$\text{Base voltamperes} = 6250 = 1 \text{ pu} \quad \text{and} \quad \text{base voltage} = 220 = 1 \text{ pu}$$

Then
$$\text{Base current} = \frac{6250}{\sqrt{3} \times 220} = 16.4 = 1 \text{ pu}$$

and
$$\text{Base reactance} = \frac{220}{16.4} = 13.4 = 1 \text{ pu}$$

so that
$$\text{Per-unit reactance} = \frac{8.4}{13.4} = 0.627 \text{ pu}$$

For the 230-V, 7.5-kVA base we obtain, from (2.6),

$$\text{Per-unit reactance} = 0.627\left(\frac{220}{230}\right)^2 \frac{7500}{6250} = 0.688 \text{ pu}$$

2.12 A three-phase, 13-kV transmission line delivers 8 MVA of load. The per-phase impedance of the line is $(0.01 + j0.05)$ pu, referred to a 13-kV, 8-MVA base. What is the voltage drop across the line?

The given base quantities yield

$$\text{Base kVA} = 8000 = 1 \text{ pu}$$

and
$$\text{Base kV} = 13 = 1 \text{ pu}$$

Then the other base quantities are

$$\text{Base current} = \frac{8000}{13\sqrt{3}} = 355.3 = 1 \text{ pu}$$

and
$$\text{Base impedance} = \frac{13,000}{355.3} = 36.6 = 1 \text{ pu}$$

From these, we find the actual values as

$$\text{Impedance} = 36.6(0.01 + j0.05) = (0.366 + j1.83) \,\Omega$$

and
$$\text{Voltage drop} = 355.3(0.366 + j1.83) = 130 + j650 = 663.1 \text{ V}$$

2.13 A portion of a power system consists of two generators in parallel, connected to a step-up transformer that links them with a 230-kV transmission line. The ratings of these components are

$$\text{Generator } G_1: \; 10 \text{ MVA, 12 percent reactance}$$

$$\text{Generator } G_2: \; 5 \text{ MVA, 8 percent reactance}$$

$$\text{Transformer: } 15 \text{ MVA, 6 percent reactance}$$

$$\text{Transmission line: } (4 + j60) \; \Omega, \; 230 \text{ kV}$$

where the percent reactances are computed on the basis of the individual component ratings. Express the reactances and the impedance in percent with 15 MVA as the base value.

Equation (2.7) gives, for generator G_1,

$$\text{Percent reactance} = 12\left(\frac{15}{10}\right) = 18 \text{ percent}$$

For generator G_2,

$$\text{Percent reactance} = 8\left(\frac{15}{5}\right) = 24 \text{ percent}$$

For the transformer,

$$\text{Percent reactance} = 6\left(\frac{15}{15}\right) = 6 \text{ percent}$$

And for the transmission line, from (2.2) and (2.7),

$$\text{Percent impedance} = (4 + j60)\frac{15 \times 10^6}{(230 \times 10^3)^2} \times 100 = (0.113 + j1.7) \text{ percent}$$

2.14　Draw an impedance diagram for the system shown in Fig. 2-4(a), expressing all values as per-unit values.

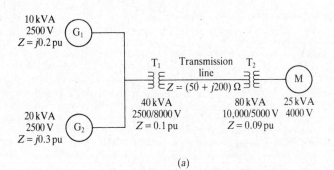

10 kVA
2500 V
$Z = j0.2$ pu
G_1

T_1　Transmission line
$Z = (50 + j200) \; \Omega$　T_2
M

20 kVA
2500 V
$Z = j0.3$ pu
G_2

40 kVA
2500/8000 V
$Z = 0.1$ pu

80 kVA
10,000/5000 V
$Z = 0.09$ pu

25 kVA
4000 V

(a)

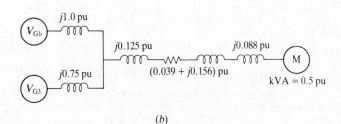

V_{G1}　$j1.0$ pu

$j0.125$ pu　　　　　　$j0.088$ pu
M

V_{G3}　$j0.75$ pu
$(0.039 + j0.156)$ pu

kVA = 0.5 pu

(b)

Fig. 2-4.

We arbitrarily choose 50 kVA to be the base kVA. Then, from (2.6), for generator G_1,

$$Z_{pu} = j0.2 \frac{(2500)^2(50)}{(2500)^2(10)} = j1.0 \text{ pu}$$

For generator G_2,

$$Z_{pu} = j0.3 \frac{(2500)^2(50)}{(2500)^2(20)} = j0.75 \text{ pu}$$

For transformer T_1,

$$Z_{pu} = j0.1 \frac{(2500)^2(50)}{(2500)^2(40)} = j0.125 \text{ pu}$$

For the transmission line,

$$Z_{pu} = (50 + j200) \frac{50{,}000}{8000^2} = 0.039 + j0.156 \text{ pu}$$

For transformer T_2,

$$Z_{pu} = j0.09 \frac{(10{,}000)^2(50)}{(8000)^2(80)} = j0.088 \text{ pu}$$

And finally for motor M,

$$kVA_{pu} = \frac{25}{50} = 0.5 \text{ pu}$$

These values produce the reactance diagram in Fig. 2-4(b).

2.15 Draw an impedance diagram for the system shown in Fig. 2-5(a), expressing all values as percent values.

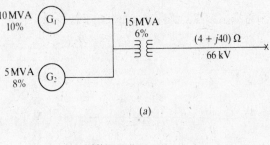

(a)

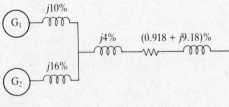

(b)

Fig. 2-5.

Let us arbitrarily choose 10 MVA to be the base MVA. Then, for generator G_1

$$\text{Percent impedance} = 10\left(\frac{10}{10}\right) = 10 \text{ percent}$$

For generator G_2,

$$\text{Percent impedance} = 8\left(\frac{10}{5}\right) = 16 \text{ percent}$$

For the transformer,

$$\text{Percent impedance} = 6\left(\frac{10}{15}\right) = 4 \text{ percent}$$

And for the transmission line,

$$\text{Percent impedance} = (4 + j40)\frac{10 \times 10^6}{(66 \times 10^3)^2} \times 100 = (0.918 + j9.18) \text{ percent}$$

These values produce Fig. 2-5(b).

2.16 Draw a per-unit reactance diagram for the system shown in Fig. 2-6(a).

We arbitrarily choose 20 MVA and 66 kV as base values. The per-unit reactance diagram is that shown as Fig. 2-6(b), where, for G_1, $X_{pu} = j0.15$ pu because its percent reactance is 15 percent with the same kVA base. Also for G_2 and G_3,

$$X_{pu} = \frac{20}{10}j0.1 = j0.2 \text{ pu}$$

For T_1 and T_2,

$$X_{pu} = \frac{20}{30}j0.15 = j0.1 \text{ pu}$$

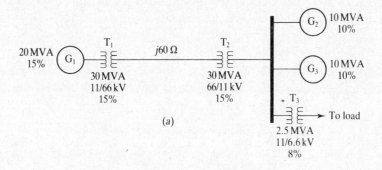

(a)

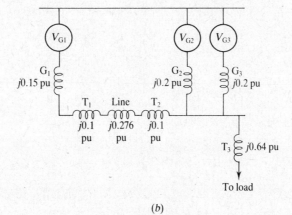

(b)

Fig. 2-6.

For T_3,

$$X_{pu} = \frac{20}{2.5} j0.08 = j0.64 \text{ pu}$$

and for the line,

$$X_{pu} = X_{line} \frac{\text{base kVA}}{(\text{base kV})^2 1000} = j60 \frac{20,000}{(66)^2(1000)} = j0.276 \text{ pu}$$

Supplementary Problems

2.17 A system operates at 220 kVA and 11 kV. Using these quantities as base values, find the base current and base impedance for the system.

Ans.   20 A; 550 Ω

2.18 Using 220 kVA and 11 kV as base values, express 138 kV, 2 MVA, 60 A, and 660 Ω as per-unit values.

Ans.   12.54 pu; 9.09 pu; 3 pu; 1.2 pu

2.19 If 25 Ω and 125 A are the base impedance and base current, respectively, for a system, find the base kVA and base voltage.

Ans.   390.625 kVA; 3125 V

2.20 The percent values of the voltage, current, impedance, and voltamperes for a given power system are 90, 30, 80, and 150 percent, respectively. The base current and base impedance are 60 A and 40 Ω, respectively. Calculate the actual values of the voltage, current, impedance, and voltamperes.

Ans.   2160 V; 18 A; 24 Ω; 5832 kVA

2.21 A single-phase transmission line supplies a reactive load at a lagging power factor. The load draws 1.2 pu current at 0.6 pu voltage while drawing 0.5 pu (true) power. If the base voltage is 20 kV and the base current is 160 A, calculate the power factor and the ohmic value of the resistance of the load.

Ans.   0.694; 43.375 Ω

2.22 The per-unit impedance of a system is 0.7 pu. The base kVA is 300 kVA, and the base voltage is 11 kV. (*a*) What is the ohmic value of the impedance? (*b*) Will this ohmic value change if 400 kVA aⅼ a ˋ kV are chosen as base values? (*c*) What is the per-unit impedance referred to the 400-kVA and 38-kV base values?

Ans.   (*a*) 282.33 Ω; (*b*) no; (*c*) 0.0782 pu

2.23 The one-line diagram for a two-generator system is shown in Fig. 2-7(*a*). Redraw the diagram to show all values as per-unit values referred to a 7000-kVA base.

Ans.   Fig. 2-7(*b*)

2.24 Redraw Fig. 2-7(*a*) to show all impedance values in ohms.

Ans.   Fig. 2-8

2.25 A 100-kVA, 20/5-kV transformer has an equivalent impedance of 10 percent. Calculate the impedance of the transformer referred to (*a*) the 20-kV side and (*b*) the 5-kV side.

Ans.   (*a*) 400 Ω; (*b*) 25 Ω

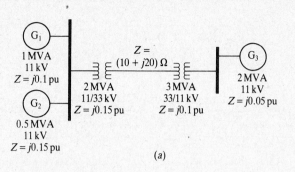

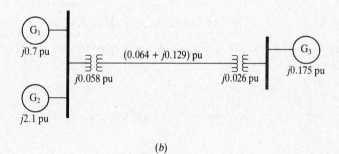

(a)

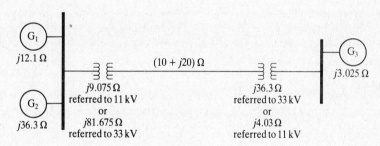

(b)

Fig. 2-7.

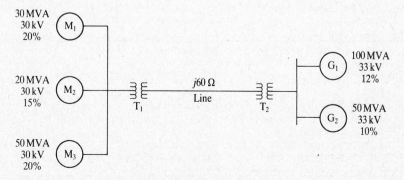

Fig. 2-8.

Fig. 2-9.

2.26 Three-phase generators G_1 and G_2 supply motor loads M_1, M_2, and M_3, as shown in Fig. 2-9. Transformers T_1 and T_2 are rated at 100 MVA and 33/110 kV, and each has a reactance of 0.08 per unit. Assuming 100 MVA and 33 kV are used as base values, obtain all the reactances as per-unit values.

Ans. Transformers, 0.08 pu; line, 0.496 pu; motors, 0.551, 0.620, and 0.331 pu

2.27 Use the results of Problem 2.26 to draw a reactance diagram for the system shown in Fig. 2-9.

Ans. Fig. 2-10

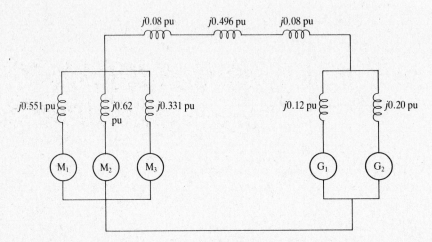

Fig. 2-10.

2.28 Three impedances, $Z_1 = 6\underline{/20°}\ \Omega$, $Z_2 = 8\underline{/40°}\ \Omega$, and $Z_3 = 10\underline{/0°}\ \Omega$, are connected in wye and are supplied by a 480-V, three-phase source. Find the line currents. Draw phasor diagrams showing all voltages and currents.

Ans. $I_a = 46.19\underline{/-50°}$ A; $I_b = 34.64\underline{/-190°}$ A; $I_c = 27.71\underline{/90°}$ A

2.29 A three-phase balanced load has a 10-Ω resistance in each of its phases. The load is supplied by a 220-V, three-phase source. Calculate the power absorbed by the load if it is connected (*a*) in wye and (*b*) in delta.

Ans. (*a*) 14.52 kW; (*b*) 14.52 kW

2.30 A three-phase, three-wire, 500-V, 60-Hz source supplies a three-phase induction motor, a wye-connected capacitor bank that draws 2 kvar per phase, and a balanced three-phase heater that draws a total of 10 kW. The induction motor is operating at its rated 75 hp and has an efficiency and power factor of 90.5 and 89.5 percent, respectively. Draw a one-line diagram for the system, and determine (*a*) the system kW, (*b*) the system kvar, and (*c*) the system kVA.

Ans. (*a*) 71.82 kW; (*b*) 24.81 kvar; (*c*) 75.98 kVA

2.31 A 440-V, three-phase source supplies a wye-connected 10-kVA load at 0.8 lagging power factor and a delta-connected, 10-kVA, unity-power-factor load. Calculate the total apparent power input to the two loads.

Ans. $18.97\underline{/18.43°}$ kVA

2.32 What is the overall power factor of the two loads of Problem 2.31? Verify that the same power factor is obtained from true power calculations.

Ans. 0.95

2.33 Calculate the line current to each of the loads of Problem 2.31.

Ans. 13.12 A

2.34 Determine the line current drawn by each load of the system of Problem 2.30.

Ans. 79.76 A; 6.93 A; 11.55 A

2.35 A balanced delta-connected load whose impedance is $45/70°\ \Omega$ per branch, a three-phase motor that draws a total of 10 kVA at 0.65 power factor lagging, and a wye-connected load whose impedance is $10\ \Omega$ (resistance) per branch are supplied from a three-phase, three-wire, 208-V, 60-Hz source. Sketch the circuit, and determine the line current to each three-phase load.

Ans. 8 A; 27.76 A; 12 A

2.36 Determine the currents I_a and I_c for the circuit in Fig. 2-11, given that $Z_1 = 20/0°\ \Omega$, $Z_2 = 14/45°\ \Omega$, $Z_3 = 14/-45°\ \Omega$, and the three-phase applied voltage is 208 V.

Ans. $15.78/-65.46°$ A; $28.70/90°$ A

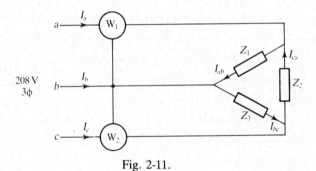

Fig. 2-11.

2.37 Find the wattmeter readings in the circuit of Fig. 2-11.

Ans. 1363.21 W, 5169.83 W

2.38 Calculate I_b for the circuit of Fig. 2-11. Determine the phasor sum of the three currents in the three phases. Explain the significance of your result.

Ans. $15.78/-114.54°$ A; $18.09/0°$ A, which is *not* equal to zero

2.39 Verify that the sum of the line currents in a delta-connected load is always zero.

2.40 Use the result of Problem 2.39 to find I_b in Fig. 2-11, and verify that the result is the same as that obtained in Problem 2.38.

Chapter 3

Transmission-Line Parameters

As noted earlier, the transmission line is one of the major components of a power system. As such, it may be represented quantitatively by a combination of three characteristics, or *parameters*: its resistance, inductance, and capacitance.

3.1 RESISTANCE

The most significant effect of the resistance of transmission-line conductors is the generation of I^2R loss in the line. The resistance also produces an IR-type voltage drop, affecting the voltage regulation of the line.

The dc resistance R of a conductor of length l and cross-sectional area A is

$$R = \rho \frac{l}{A} \quad \text{(in ohms)} \tag{3.1}$$

where ρ is the *resistivity* of the material of the conductor in ohm-meters. The dc resistance of a conductor is affected only by the operating temperature, and it increases linearly with the temperature. However, when a conductor is transmitting alternating current, the current-density distribution across the conductor cross section is nonuniform and is a function of the ac frequency. This phenomenon, known as the *skin effect,* causes the ac resistance to be greater than the dc resistance. At 60 Hz, the ac resistance of a transmission-line conductor may be 5 to 10 percent higher than its dc resistance.

The temperature dependence of resistance is quantified by the relation

$$R_2 = R_1[1 + \alpha(T_2 - T_1)] \tag{3.2}$$

where R_1 and R_2 are the resistances at temperatures T_1 and T_2, respectively, and α is called the *temperature coefficient of resistance*. The resistivities and temperature coefficients of several metals are given in Table 3-1.

TABLE 3-1 Resistivities and Temperature Coefficients of Resistance

Material	Resistivity ρ at 20°C $\mu\Omega \cdot$ cm	Temperature coefficient α at 20°C, °C^{-1}
Aluminum	2.83	0.0039
Brass	6.4–8.4	0.0020
Copper		
Hard-drawn	1.77	0.00382
Annealed	1.72	0.00393
Iron	10.0	0.0050
Silver	1.59	0.0038
Steel	12–88	0.001–0.005

Long transmission lines may involve shunt resistances (or conductances) in addition to series resistances.

3.2 INDUCTANCE

Two-Wire, Single-Phase Line

The inductance per conductor of a two-wire, single-phase transmission line is given by

$$L_1 = \frac{\mu_0}{8\pi} \left(1 + 4\ln\frac{D}{r}\right) \quad \text{(in henrys per meter)} \tag{3.3}$$

where $\mu_0 = 4\pi \times 10^{-7}\,\text{H/m}$ (the permeability of free space), D is the distance between the centers of the conductors, and r is the radius of the conductors. The total, or loop, inductance is then

$$L = 2L_1 = \frac{\mu_0}{4\pi}\left(1 + 4\ln\frac{D}{r}\right)$$

$$= \left(1 + 4\ln\frac{D}{r}\right) \times 10^{-7} \quad \text{H/m} \tag{3.4}$$

Since $\ln e^{1/4} = 1/4$, this last equation may also be written as

$$L = 4 \times 10^{-7} \ln\frac{D}{r'} \quad \text{H/m} \tag{3.5}$$

where $r' = re^{-1/4}$ is known as the *geometric mean radius* (GMR) of the conductor.

Of the two terms in (3.3), the first represents the internal inductance of the solid conductor, and the second term is due to fluxes external to the conductor. In (3.5), the conductor is replaced by an equivalent thin-walled, hollow conductor of radius r' having no internal flux linkage and hence no internal inductance.

Three-Wire, Three-Phase Line

The per-phase (or line-to-neutral) inductance of a three-phase transmission line with equilaterally spaced conductors is

$$L = \frac{\mu_0}{8\pi}\left(1 + 4\ln\frac{D}{r}\right)$$

$$= 2\left(\frac{1}{4} + \ln\frac{D}{r}\right) \times 10^{-7} \quad \text{H/m} \tag{3.6}$$

where r is the conductor radius and D is the spacing between conductors. In practice, the three conductors of a three-phase line are seldom equilaterally spaced. The usual nonsymmetrical spacing produces unequal inductances in the three phases, leading to unequal voltage drops and an imbalance in the line. To offset this imbalance, the positions of the conductors are interchanged at regular intervals along the line. This practice is known as *transposition* and is illustrated in Fig. 3-1, which also shows the unequal spacings between conductors. The average per-phase inductance for a transposed line is still given by (3.6), except that the spacing D in the equation is replaced by the

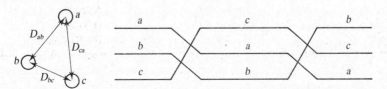

Fig. 3-1.

equivalent spacing D_e obtained from

$$D_e = (D_{ab}D_{bc}D_{ca})^{1/3} \tag{3.7}$$

where the distances D_{ab}, D_{bc}, and D_{ca} are as shown in Fig. 3-1.

Composite Conductors

These expressions for the line inductance must be modified for application to a transmission line that consists of composite conductors. In particular, let a single-phase line consist of two composite conductors, as shown in Fig. 3-2. Conductor X is composed of n identical and parallel filaments, each of which carries the current I/n. Conductor Y, which is the return circuit for the current in conductor X, is composed of m identical and parallel filaments, each of which carries the current $-I/m$. Distances between pairs of elements are designated by D with appropriate subscripts. The inductance L_X of conductor X then may be shown to be

$$L_X = 2 \times 10^{-7} \ln \frac{\sqrt[mn]{(D_{aa'}D_{ab'}D_{ac'}\cdots D_{am})(D_{ba'}D_{bb'}D_{bc'}\cdots D_{bm})\cdots(D_{na'}D_{nb'}D_{nc'}\cdots D_{nm})}}{\sqrt[n^2]{(D_{aa}D_{ab}D_{ac}\cdots D_{an})(D_{ba}D_{bb}D_{bc}\cdots D_{bn})\cdots(D_{na}D_{nb}D_{nc}\cdots D_{nm})}}$$

$$\text{(in henrys per meter)} \quad (3.8)$$

where $D_{kk} = r'_k = r_k e^{-1/4}$ is the geometric mean radius (GMR) of the kth conductor. [The GMR is defined just below (3.5).] Notice that the numerator in (3.8) involves the mnth root of the product of mn terms; each of those terms is the distance from one of the n filaments of conductor X to one of the m filaments of conductor Y, and there are a total of mn distances. The mnth root of the product of mn distances is called a *geometric mean distance*. For two conductors X and Y, as in Fig. 3-2, it is called the *mutual* geometric mean distance between them and abbreviated D_m or GMD.

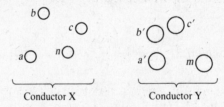

Conductor X   Conductor Y

Fig. 3-2.

The n^2 root of the product of n^2 distances occurring in the denominator of (3.8) is abbreviated D_s and called the *self* GMD of conductor X. In like manner, r' for a separate filament or wire is often called its self GMD. The self GMD is also sometimes loosely termed the *geometric mean radius* and abbreviated GMR.

In terms of D_m and D_s, (3.8) becomes

$$L_X = 2 \times 10^{-7} \ln \frac{D_m}{D_s} \quad \text{H/m} \tag{3.9}$$

We determine the inductance L_Y of conductor Y in a similar manner, and the total line inductance becomes

$$L = L_X + L_Y \tag{3.10}$$

Double-Circuit, Three-Phase Line

The per-phase inductance of a double-circuit, three-phase, transposed transmission line (Fig. 3-3) is given by

$$L = 2 \times 10^{-7} \ln \frac{\text{GMD}}{\text{GMR}} \quad \text{H/m} \tag{3.11}$$

In terms of the symbols of Fig. 3-3, which shows a transposed three-phase line, (3.11) may be written as

$$L = 2 \times 10^{-7} \ln \frac{\sqrt[6]{2}\sqrt{D}\sqrt[3]{G}}{\sqrt{r'}\sqrt[3]{F}} \tag{3.12}$$

where r' is the GMR of the conductor.

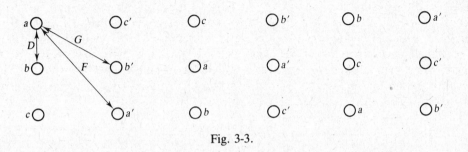

Fig. 3-3.

3.3 CAPACITANCE

The shunt capacitance per unit length of a single-phase, two-wire transmission line is given by

$$C = \frac{\pi \epsilon_0}{\ln (D/r)} \quad \text{(in farads per meter)} \tag{3.13}$$

where ϵ_0 is the permittivity of free space and the other symbols are as defined for (3.3). For a three-phase line with equilaterally spaced conductors, the per-phase (or line-to-neutral) capacitance is

$$C = \frac{2\pi \epsilon_0}{\ln (D/r)} \quad \text{F/m} \tag{3.14}$$

To account for the actual unequal spacings between conductors and the transposition of the line, D in (3.14) is replaced with D_e of (3.7), as is done in computing the inductance of a transposed line.

For the double-circuit transmission line of Fig. 3-3, the per-phase capacitance is given by

$$C = \frac{4\pi \epsilon_0}{\ln [\sqrt[3]{2}\,(D/r)(G/F)^{2/3}]} \quad \text{F/m} \tag{3.15}$$

The capacitance of an overhead transmission line is affected by the ground, which distorts its electric field. The effect of the earth is simulated by assuming the existence of mirror-image conductors, as far below ground level as the transmission line is above it (Fig. 3-4). The image conductors carry charges with polarities opposite those of the real conductors, as shown. Now the capacitance to neutral is given by

$$C_n = \frac{2\pi \epsilon_0}{\ln (D_e/r) - \ln (\sqrt[3]{H_{ab}H_{bc}H_{ca}}/\sqrt[3]{H_aH_bH_c})} \quad \text{F/m} \tag{3.16}$$

where D_e is given by (3.7), the H's are defined in Fig. 3-4, and r is the conductor radius.

Using the concept of the GMD, we may write the capacitance to neutral of a nonsymmetrical three-phase double-circuit line as

$$C_n = \frac{2\pi \epsilon_0}{\ln (\text{GMD}/\text{GMR})} = \frac{2\pi \epsilon_0}{\ln (D_m/D_s)} \quad \text{F/m} \tag{3.17}$$

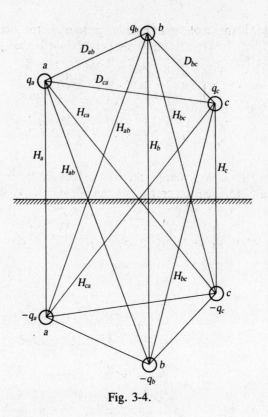

Fig. 3-4.

Substituting the numerical value for ϵ_0 in (3.17) yields

$$C_n = \frac{10^{-9}}{18 \ln{(D_m/D_s)}} \quad \text{F/m} \tag{3.18}$$

Solved Problems

3.1 Determine the resistance of a 10-km-long solid cylindrical aluminum conductor with a diameter of 250 mils, at (a) 20°C and (b) 120°C.

To find the cross-sectional area of the conductor, we note that

$$250 \text{ mils} = 0.25 \text{ in} = 0.635 \text{ cm}$$

so

$$A = \frac{\pi}{4}(0.635)^2 = 0.317 \text{ cm}^2$$

Also, from Table 3-1, $\rho = 2.83 \; \mu\Omega \cdot \text{cm}$ and $\alpha = 0.0039°\text{C}^{-1}$ at 20°C.

(a) At 20°C, (3.1) yields

$$R_{20} = \rho \frac{l}{A} = 2.83 \times 10^{-8} \times \frac{10 \times 10^3}{0.317 \times 10^{-4}} = 8.93 \; \Omega$$

(b) At 120°C, (3.2) yields

$$R_{120} = R_{20}[1 + \alpha(120 - 20)] = 8.93(1 + 0.0039 \times 100) = 12.41 \; \Omega$$

3.2 A transmission-line cable consists of 19 strands of identical copper conductors, each 1.5 mm in diameter. The length of the cable is 2 km but, because of the twist of the strands, the actual length of each conductor is increased by 5 percent. What is the resistance of the cable? Take the resistivity of copper to be $1.72 \times 10^{-8}\,\Omega \cdot m$.

Allowing for twist, we find that $l = (1.05)(2000) = 2100\,m$. The cross-sectional area of all 19 strands is $19(\pi/4)(1.5 \times 10^{-3})^2 = 33.576 \times 10^{-6}\,m^2$. Then, from (3.1),

$$R = \frac{\rho l}{A} = \frac{1.72 \times 10^{-8} \times 2100}{33.576 \times 10^{-6}} = 1.076\,\Omega$$

3.3   The variation of resistance with temperature is expressed by the temperature coefficient of resistance α. Explicitly, the resistance R_T at a temperature $T°C$ is related to the resistance R_0 at 0°C by $R_T = R_0(1 + \alpha_0 T)$, where α_0 is the temperature coefficient at 0°C. This relation is depicted for copper in Fig. 3-5, which also shows the inferred absolute zero for copper. Using Fig. 3-5, find the resistance of a copper wire at −20°C if its resistance at 0°C is 20 Ω.

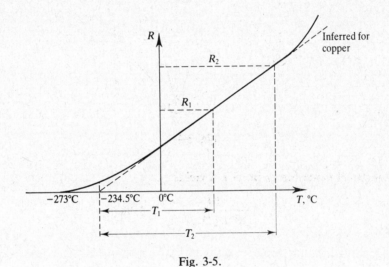

Fig. 3-5.

From Fig. 3-5, we have $\alpha_0 = 1/234.5$. From the given data,

$$R_2 = R_0(1 + \alpha_0 T_2) = 20\left(1 + \frac{-20}{234.5}\right) = 18.29\,\Omega$$

3.4   A sample of copper wire has a resistance of 50 Ω at 10°C. What must be the maximum operating temperature of the wire if its resistance is to increase by at most 10 percent? Take the temperature coefficient at 10°C to be $\alpha = 0.00409°C^{-1}$

Here we have $R_1 = 50\,\Omega$ and $R_2 = 50 + 0.1 \times 50 = 55\,\Omega$. Also, $T_1 = 10°C$, and we require T_2. From (3.2) we obtain

$$55 = 50[1 + 0.00409(T_2 - 10)] \quad \text{or} \quad T_2 = 34.45°C$$

3.5   The per-phase line loss in a 40-km-long transmission line is not to exceed 60 kW while it is delivering 100 A per phase. If the resistivity of the conductor material is $1.72 \times 10^{-8}\,\Omega \cdot m$, determine the required conductor diameter.

The line loss is to be, at most,

$$I^2 R = (100)^2 R = 60 \times 10^3$$

from which we find $R = 6$. Substituting that value into (3.1), solving the result for A, and substituting $A = \pi D^2/4$ yield

$$\frac{\pi}{4}D^2 = \frac{(1.72 \times 10^{-8})(40 \times 10^3)}{6}$$

from which $D = 1.208$ cm.

3.6 A coaxial cable has an inner conductor of radius r_1 and a hollow outer conductor of radius r_2 (Fig. 3-6). The outer conductor is of negligible thickness. Determine the inductance per unit length of the cable by finding the energy stored in the magnetic field of the cable and equating it to the energy stored in the cable inductance. Assume a uniform current-density distribution over the conductor cross section.

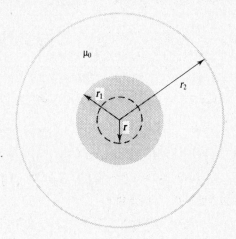

Fig. 3-6.

From Ampere's law, the magnetic field intensities inside and outside the inner conductor are, respectively,

$$H_{\phi i} = \frac{Ir}{2\pi r^2} \qquad \text{for } 0 < r < r_1 \tag{1}$$

$$H_{\phi e} = \frac{I}{2\pi r} \qquad \text{for } r_1 < r < r_2 \tag{2}$$

Also, $W_m = \frac{1}{2}\int_v \mathbf{B} \cdot \mathbf{H} \, dv$; since $\mathbf{B} = \mu_0 \mathbf{H}$, this becomes, for a unit length,

$$W_m = \frac{1}{2}\mu_0\left(\int_0^{r_1} H_{\phi i}^2 2\pi r \, dr + \int_{r_1}^{r_2} H_{\phi e}^2 2\pi r \, dr\right) \tag{3}$$

Substituting (1) and (2) in (3) yields

$$W_m = \frac{\mu_0 I^2}{4\pi}\left(\frac{1}{4} + \ln\frac{r_2}{r_1}\right)$$

But $W_m = \frac{1}{2}LI^2$, so that

$$L = \frac{\mu_0}{2\pi}\left(\frac{1}{4} + \ln\frac{r_2}{r_1}\right) \qquad \text{H/m}$$

3.7 A single-phase, two-wire transmission line, 15 km long, is made up of round conductors, each

0.8 cm in diameter, separated from each other by 40 cm. Calculate the equivalent diameter of a fictitious hollow, thin-walled conductor having the same inductance as the original line. What is the value of this inductance?

The fictitious conductor is one whose radius is r' and whose diameter is therefore

$$2r' = re^{-1/4} = 0.8 \times 0.7788 = 0.623 \text{ cm}$$

Using (3.5), we find that the inductance of 15 km of such a conductor is

$$L = 15 \times 10^3 \times 4 \times 10^{-7} \ln \frac{40}{0.5 \times 0.623} = 29.13 \text{ mH}$$

3.8 A single-circuit, three-phase, 60-Hz transmission line consists of three conductors arranged as shown in Fig. 3-7. If the conductors are the same as that in Problem 3.1, find the inductive reactance of the line per kilometer per phase.

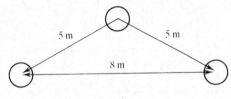

5 m   5 m

8 m

Fig. 3-7.

From (3.7),

$$D_e = (5 \times 5 \times 8)^{1/3} = 5.848 \text{ m}$$

From Problem 3.1, $r = \frac{1}{2} \times 0.635 \times 10^{-2}$ m, so that

$$\frac{D_e}{r} = \frac{5.848 \times 2 \times 10^2}{0.635} = 1841.9$$

and $\ln(D_e/r) = 7.52$. Hence, from (3.6) we have, for each kilometer of length,

$$L = 2(\tfrac{1}{4} + 7.52) \times 10^{-7} \times 10^3 = 1.554 \text{ mH/km}$$

The inductive reactance per kilometer is then

$$X_L = \omega L = 377 \times 1.554 \times 10^{-3} = 0.5858 \ \Omega$$

3.9 Calculate the capacitance and capacitive reactance (at 60 Hz) of the transmission line of Problem 3.7.

For air, $\epsilon_0 = 10^{-9}/36\pi$ F/m. Thus, from (3.13),

$$C = 15 \times 10^3 \frac{\pi(10^{-9}/36\pi)}{\ln(40/0.4)} = 0.0904 \ \mu\text{F}$$

and

$$X_C = \frac{1}{\omega C} = \frac{1}{377(0.0904 \times 10^{-6})} = 29.34 \text{ k}\Omega$$

3.10 What is the capacitive reactance per kilometer of the three-phase transmission line of Problem 3.8?

From (3.14) with $\epsilon_0 = 10^{-9}/36\pi$ F/m,

$$C = \frac{2\pi \times 10^{-9}/36\pi}{7.52} \times 10^3 = 7.387 \times 10^{-9} \text{ F/km}$$

Hence the capacitive reactance per kilometer is

$$X_C = \frac{1}{\omega C} = \frac{10^9}{377 \times 7.387} = 0.36 \times 10^6 \, \Omega/\text{km}$$

3.11 Find the inductance per unit length of the single-phase line shown in Fig. 3-8. Conductors a, b, and c are of 0.2 cm radius, and conductors d and e are of 0.4 cm radius.

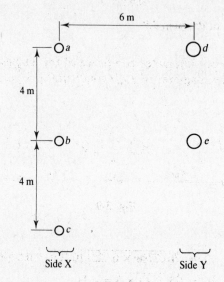

Fig. 3-8.

Because the line is not symmetrical, we use (*3.8*) or (*3.9*). The GMD between sides X and Y is

$$D_m = \sqrt[6]{D_{ad}D_{ae}D_{bd}D_{be}D_{cd}D_{ce}}$$

where

$$D_{ad} = D_{be} = 6 \, \text{m}$$

$$D_{ae} = D_{bd} = D_{ce} = \sqrt{4^2 + 6^2} = 7.21 \, \text{m}$$

and

$$D_{cd} = \sqrt{6^2 + 8^2} = 10 \, \text{m}$$

Hence

$$D_m = \sqrt[6]{6 \times 7.21 \times 7.21 \times 6 \times 10 \times 7.21} = 7.16 \, \text{m}$$

The GMR for side X is, with $D_{kk} = r_k e^{-1/4} = 0.7788 r_k$,

$$D_{sX} = \sqrt[9]{D_{aa}D_{ab}D_{ac}D_{ba}D_{bb}D_{bc}D_{ca}D_{cb}D_{cc}}$$

$$= \sqrt[9]{(0.20 \times 10^{-2} \times 0.7788)^3 \times 4 \times 8 \times 4 \times 4 \times 8 \times 4} = 0.341 \, \text{m}$$

and that for side Y is

$$D_{sY} = \sqrt[4]{(0.4 \times 0.7788 \times 10^{-2})^2 \times 4^2} = 0.112 \, \text{m}$$

Now, from (*3.9*),

$$L_X = 2 \times 10^{-7} \ln \frac{7.16}{0.341} = 6.09 \times 10^{-7} \, \text{H/m}$$

and

$$L_Y = 2 \times 10^{-7} \ln \frac{7.16}{0.112} = 8.31 \times 10^{-7} \, \text{H/m}$$

Then

$$L = L_X + L_Y = 14.4 \times 10^{-7} \, \text{H/m}$$

3.12 Verify the result of Problem 3.8 by applying the concept of the GMR and GMD.

From Problem 3.1, the diameter of the conductor is 0.635 cm. Hence,

$$\text{GMR} = D_s = \frac{0.7788 \times 0.635}{2 \times 100} = 0.002473 \text{ m}$$

and

$$\text{GMD} = D_m = \sqrt[3]{5 \times 8 \times 5} = 5.848 \text{ m}$$

Thus, the inductance per kilometer is, from (3.9),

$$L = 2 \times 10^{-7} \ln \frac{5.848}{0.002473} \times 1000 = 1.554 \text{ mH/km}$$

which agrees with the result of Problem 3.8.

3.13 Calculate the capacitance per kilometer per phase of the single-circuit, two-bundle conductor line shown in Fig. 3-9. The diameter of each conductor is 5 cm.

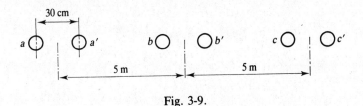

Fig. 3-9.

We have

$$D_s = \sqrt[4]{(0.7788 \times 0.025)^2 (0.30)^2} = 0.07643 \text{ m}$$

$$D_m = \sqrt[3]{5 \times 10 \times 5} = 6.3 \text{ m}$$

Hence, from (3.18),

$$C_n = \frac{10^{-9} \times 1000}{18 \ln (6.3/0.07643)} = 0.0126 \ \mu\text{F/km}$$

Supplementary Problems

3.14 A single-phase transmission line, 50 km long, is made up of a hard-drawn copper conductor 500 mils in diameter. Using data from Table 3-1, find the loop resistance at 20°C.

Ans.   0.1394 Ω

3.15 Determine the resistance of the line of Problem 3.14 at 80°C.

Ans.   0.1714 Ω

3.16 A transmission-line conductor has a resistance of 7 Ω at 0°C. Calculate the temperature coefficient of the conductor metal at 20°C if its resistance increases to 7.8 Ω at 20°C.

Ans.   0.00513°C⁻¹

3.17 The resistance of a transmission line is 25 Ω at 15°C and increases by 10 percent when the operating temperature increases to 50°C. At what temperature is its resistance 30 Ω, if the temperature coefficient is assumed to remain constant.

Ans.   65°C

3.18 The conductors of a three-phase transmission line are arranged in the form of an equilateral triangle with sides of 6 m each. If the conductors are 500 mils in diameter and the line is 25 km long, what is its inductance per phase?

Ans. 35.5 mH

3.19 Let $D_{ab} = c$, $D_{bc} = a$, and $D_{ca} = b$ in the three-phase transposed transmission line shown in Fig. 3-1. Obtain an expression for the inductance of phase a for a 1-m-long line. The conductor radius is r.

Ans. $\left(\dfrac{1}{2} + 2\ln\dfrac{\sqrt{bc}}{r} + j\sqrt{3}\ln\dfrac{c}{b}\right) \times 10^{-7}\,\text{H}$

3.20 A single-phase, 10-km transmission line has 16.65 mH total inductance. If the distance between the conductors is 1.0 m, what is the conductor diameter?

Ans. 2.0 cm

3.21 Determine the geometric mean radius of the conductors of Problem 3.20.

Ans. 1.558 cm

3.22 A double-circuit, single-phase transmission line is shown in Fig. 3-10. Obtain an expression for the inductance per meter of each conductor.

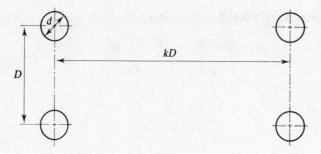

Fig. 3-10.

Ans. $\left(\dfrac{1}{2} + 2\ln\dfrac{2kD}{d}\sqrt{1 + k^2}\right) \times 10^{-7}\,\text{H/m}$

3.23 Calculate the per-phase capacitance of the transmission line described in Problem 3.18.

Ans. 0.203 μF

3.24 A single-phase transmission line is h meters above the ground and consists of conductors of radius r meters separated from each other by a distance of d meters. Obtain an expression for the capacitance per meter between conductors, including the effect of the ground.

Ans. $\pi\epsilon_0/\ln\left[d/r\sqrt{1 + (d/2h)^2}\right]$ F/m

3.25 Find the GMR of a bundle of two conductors separated by a distance d, each having a GMR of D_s.

Ans. $\sqrt{D_s d}$

3.26 Rework Problem 3.25 for a bundle of four conductors located at the four corners of a square of side d. The GMR of each conductor is D_s.

Ans. $1.09\,(D_s d^3)^{1/4}$

3.27   Calculate the inductance per kilometer per phase of the line shown in Fig. 3-9.

Ans.   0.882 mH/km

3.28   A double-circuit, three-phase, transposed transmission line is shown in Fig. 3-11. The radius of each conductor is 1.25 cm. Calculate the inductance per kilometer per phase.

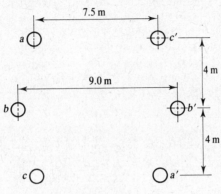

Fig. 3-11.

Ans.   0.607 mH/km

3.29   Find the capacitance to neutral of the line of Problem 3.28.

Ans.   0.019 μF/km

Chapter 4

Transmission-Line Calculations

Transmission lines physically integrate the output of generating plants and the requirements of customers by providing pathways for the flow of energy among the various circuits in an electric power system. For our purposes here, we consider a transmission line to have a sending end and a receiving end, and to have a series resistance and inductance and a shunt capacitance and conductance as primary parameters. In addition, we classify transmission lines as short, medium, and long. In a short line, the shunt effects (conductance and capacitance) are neglected; this approximation is considered valid for lines up to 80 km long. In a medium line, the shunt capacitances are lumped at a few predetermined locations along the line; medium lines generally range from 80 to 240 km in length. Lines longer than 240 km are considered to be long lines and to have uniformly distributed parameters.

In Chapter 3 we discussed the three most important parameters of transmission lines. In this chapter we discuss the effect of those parameters on the operation and performance of transmission lines. In particular, we evaluate the losses, efficiency, and voltage regulation of transmission lines and then determine the consequences of such performance characteristics on the operation of a power system.

4.1 TRANSMISSION-LINE REPRESENTATION

To facilitate performance calculations relating to a transmission line, the line is approximated as a series–parallel interconnection of the relevant parameters. A short transmission line, for which the shunt effects may be neglected, is represented by a lumped resistance in series with a lumped inductance. A medium-length line is represented by lumped shunt capacitors located at predetermined points along an RL series circuit. (In practice, the entire capacitive effect in a medium-length line may be represented by only one or two lumped capacitors.) Finally, a long transmission line is represented by uniformly distributed parameters. Furthermore, the shunt branch of a long line consists of both capacitances and conductances distributed uniformly along the line.

4.2 SHORT TRANSMISSION LINE

The *short transmission line* is represented by the lumped parameters R and L, as shown in Fig. 4-1. Notice that R is the resistance (per phase) and L is the inductance (per phase) of the *entire line* (even though we computed transmission-line parameters per unit length of line in Chapter 3). The line is shown to have two ends: the sending end (designated by the subscript S) at the generator, and the receiving end (designated R) at the load. Quantities of significance here are the voltage regulation and efficiency of transmission. These quantities are defined as follows for lines of all lengths:

$$\text{Percent voltage regulations} = \frac{|V_{R(\text{no load})}| - |V_{R(\text{load})}|}{|V_{R(\text{load})}|} \times 100 \qquad (4.1)$$

$$\text{Efficiency of transmission} = \frac{\text{power at receiving end}}{\text{power at sending end}} = \frac{P_R}{P_S} \qquad (4.2)$$

where V_R is the receiving-end voltage.

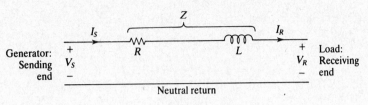

Fig. 4-1.

4.3 MEDIUM-LENGTH TRANSMISSION LINE

In a *medium-length transmission line* the shunt effect due to the line capacitance is not negligible. Two representations for such a line are shown in Figs. 4-2 and 4-3; they are known as the *nominal*-II *circuit* and the *nominal-T circuit* of the transmission line, respectively. The figures also show the phasor diagrams for lagging power-factor conditions. These diagrams are of help in understanding the mutual relationships between currents and voltages along the line.

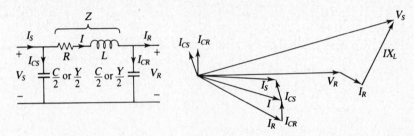

Fig. 4-2.

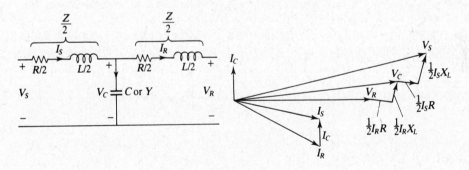

Fig. 4-3.

4.4 LONG TRANSMISSION LINE

The parameters of a *long line* are considered to be distributed over the entire length of the line. One phase (with return through neutral) of a long line, of length $\mathscr{L}$, is shown in Fig. 4-4. The voltage V at any point along this line is given by

$$\frac{d^2V}{dx^2} = \gamma^2 V \tag{4.3}$$

where $\gamma = \sqrt{yz}$, y is the shunt admittance per unit length of the line, z is the series impedance per unit length, and γ is known as the *propagation constant*. A solution to (4.3) is

$$V = \tfrac{1}{2}V_R(e^{\gamma x} + e^{-\gamma x}) + \tfrac{1}{2}I_R Z_c(e^{\gamma x} - e^{-\gamma x}) \tag{4.4}$$

where $Z_c = \sqrt{z/y}$ is called the *characteristic impedance* of the line. The current I at any point along the line is given by

$$I = \frac{1}{2}\frac{V_R}{Z_c}(e^{\gamma x} - e^{-\gamma x}) + \tfrac{1}{2}I_R(e^{\gamma x} + e^{-\gamma x}) \tag{4.5}$$

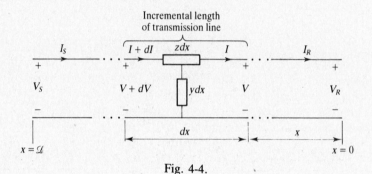

Fig. 4-4.

Equations (4.4) and (4.5) may be expressed in terms of hyperbolic functions as

$$V = V_R \cosh \gamma x + I_R Z_c \sinh \gamma x \tag{4.6}$$

$$I = V_R \frac{\sinh \gamma x}{Z_c} + I_R \cosh \gamma x \tag{4.7}$$

Since $V = V_S$ and $I = I_S$ at $x = \mathscr{L}$, (4.6) and (4.7) become, at the sending end,

$$V_S = V_R \cosh \gamma \mathscr{L} + I_R Z_c \sinh \gamma \mathscr{L} \tag{4.8}$$

$$I_S = V_R \frac{\sinh \gamma \mathscr{L}}{Z_c} + I_R \cosh \gamma \mathscr{L} \tag{4.9}$$

The following relationships are often useful in numerical computations involving (4.6) to (4.9):

$$\gamma = \alpha + j\beta$$
$$\cosh \gamma \mathscr{L} = \cosh (\alpha \mathscr{L} + j\beta \mathscr{L}) = \cosh \alpha \mathscr{L} \cos \beta \mathscr{L} + j \sinh \alpha \mathscr{L} \sin \beta \mathscr{L}$$
$$\sinh \gamma \mathscr{L} = \sinh (\alpha \mathscr{L} + j\beta \mathscr{L}) = \sinh \alpha \mathscr{L} \cos \beta \mathscr{L} + j \cosh \alpha \mathscr{L} \sin \beta \mathscr{L}$$
$$\cosh \gamma \mathscr{L} = 1 + \frac{(\gamma \mathscr{L})^2}{2!} + \frac{(\gamma \mathscr{L})^4}{4!} + \cdots \approx 1 + \tfrac{1}{2}yz \tag{4.10}$$
$$\sinh \gamma \mathscr{L} = \gamma \mathscr{L} + \frac{(\gamma \mathscr{L})}{3!} + \frac{(\gamma \mathscr{L})^5}{5!} + \cdots \approx \sqrt{yz}\,(1 + \tfrac{1}{6}yz)$$

4.5 THE TRANSMISSION LINE AS A TWO-PORT NETWORK

In preceding sections we found that, when a transmission line is represented by its equivalent circuit, we can express the sending-end voltage and current in terms of the receiving-end voltage and current and the line parameters. In general, a transmission line may be viewed as a four-terminal network, as shown in Fig. 4-5, such that the terminal voltages and currents are related by

$$V_S = AV_R + BI_R \tag{4.11}$$

$$I_S = CV_R + DI_R \tag{4.12}$$

where the constants A, B, C, and D are called the *generalized circuit constants* or *ABCD* constants and are, in general, complex. By reciprocity, they are related to each other as follows:

$$AD - BC = 1 \qquad (4.13)$$

A transmission line of any length can be represented by the four-terminal network of Fig. 4-5 with *ABCD* constants as given in Table 4-1.

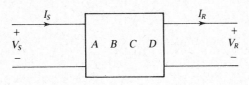

Fig. 4-5.

TABLE 4-1   ABCD Constants for Transmission Lines (per Phase)

Line length	Equivalent circuit	A	B	C	D
Short	Series impedance, Fig. 4-1	1	Z	0	1
Medium	Nominal II, Fig. 4-2	$1 + \frac{1}{2}YZ$	Z	$Y(1 + \frac{1}{4}YZ)$	$1 + \frac{1}{2}YZ$
	Nominal T, Fig. 4-3	$1 + \frac{1}{2}YZ$	$Z(1 + \frac{1}{4}YZ)$	Y	$1 + \frac{1}{2}YZ$
Long	Distributed parameters, Fig. 4-4	$\cosh \gamma \mathscr{L}$	$Z_c \sinh \gamma \mathscr{L}$	$(\sinh \gamma \mathscr{L})/Z_c$	$\cosh \gamma \mathscr{L}$

4.6 POWER FLOW ON TRANSMISSION LINES

The power flow at any point on a transmission line can conveniently be calculated in terms of the *ABCD* constants. Since these constants are generally complex, we let

$$A = |A| \underline{/\alpha} \qquad \text{and} \qquad B = |B| \underline{/\beta} \qquad (4.14)$$

Choosing V_R as the reference phasor, we assume that

$$V_R = |V_R| \underline{/0°} \qquad \text{and} \qquad V_S = |V_S| \underline{/\delta} \qquad (4.15)$$

Then, from (4.11) we obtain

$$I_R = \frac{|V_S|}{|B|} \underline{/\delta - \beta} - \frac{|A||V_R|}{|B|} \underline{/\alpha - \beta} \qquad (4.16)$$

The complex power $V_R I_R^*$ at the receiving end is thus given by

$$P_R + jQ_R = \frac{|V_R||V_S|}{|B|} \underline{/\beta - \delta} - \frac{|A||V_R|^2}{|B|} \underline{/\beta - \alpha} \qquad (4.17)$$

so that

$$P_R = \frac{|V_R||V_S|}{|B|} \cos(\beta - \delta) - \frac{|A||V_R|^2}{|B|} \cos(\beta - \alpha) \qquad (4.18)$$

$$Q_R = \frac{|V_R||V_S|}{|B|} \sin(\beta - \delta) - \frac{|A||V_R|^2}{|B|} \sin(\beta - \alpha) \qquad (4.19)$$

4.7 TRAVELING WAVES ON TRANSMISSION LINES

On a long transmission line such as that in Fig. 4-4, the voltage V and current I everywhere along the line satisfy a relation called the *wave equation*. For a lossless transmission line, such that z and y in Fig. 4-4 are purely reactive, the wave equation may be written as

$$\frac{1}{LC}\frac{\partial^2 V}{\partial x^2} = \frac{\partial^2 V}{\partial t^2} \tag{4.20}$$

or

$$\frac{\partial^2 I}{\partial x^2} = LC\frac{\partial^2 I}{\partial t^2} \tag{4.21}$$

Solutions to (4.20) and (4.21) take the forms

$$V(x, t) = V^+\left(t - \frac{x}{u}\right) + V^-\left(t + \frac{x}{u}\right) \tag{4.22}$$

and

$$I(x, t) = I^+\left(t - \frac{x}{u}\right) + I^-\left(t + \frac{x}{u}\right) \tag{4.23}$$

where

$$u = \frac{1}{\sqrt{LC}} \tag{4.24}$$

and the superscripts $+$ and $-$ denote, respectively, waves traveling in the $+x$ and $-x$ directions along the transmission line. That the solutions (4.22) and (4.23) do indeed represent traveling waves is indicated by the arguments, in which u has the dimension meters per second. A wave such as $V^+(t - x/u)$ that is traveling in the positive x direction is called a *forward-traveling* wave, and one that is moving in the negative x direction is a *backward-traveling* wave.

It may be verified from (4.20) through (4.24) that

$$\frac{V^+}{I^+} = \sqrt{\frac{L}{C}} \tag{4.25}$$

and

$$\frac{V^-}{I^-} = -\sqrt{\frac{L}{C}} \tag{4.26}$$

The ratio $\sqrt{L/C}$ has the dimension ohms and is called the *characteristic impedance* Z_c of the line. Recall from Section 4.4 that $Z_c = \sqrt{z/y}$ for a lossy line, whereas here the line is lossless and the characteristic impedance is purely resistive. We may thus call it the *characteristic resistance* R_c and write

$$Z_c = \sqrt{\frac{L}{C}} = R_c \tag{4.27}$$

for a lossless line. In terms of R_c, (4.23) becomes

$$I(x, t) = \frac{1}{R_c}V^+\left(t - \frac{x}{u}\right) - \frac{1}{R_c}V^-\left(t + \frac{x}{u}\right) \tag{4.28}$$

Figure 4-6 shows a transmission line of total length $\mathcal{L}$ that terminates in a resistance R_L and is driven by a pulse voltage source having an open-circuit voltage waveform $V_S(t)$ as shown and an internal resistance R_S. To determine the terminal voltages $V(0, t)$ and $V(\mathcal{L}, t)$ and terminal currents $I(0, t)$ and $I(\mathcal{L}, t)$ as functions of time we consider the portion of the line at the load (Fig. 4-7). At $x = \mathcal{L}$, we must have

$$V(\mathcal{L}, t) = R_L I(\mathcal{L}, t) \tag{4.29}$$

Equation (4.29) requires the existence of forward- and backward-traveling waves at $x = \mathcal{L}$. If only forward-traveling waves exist at the load, then

$$V^+\left(t - \frac{\mathscr{L}}{u}\right) = R_c I^+\left(t - \frac{\mathscr{L}}{u}\right) \tag{4.30}$$

and if only backward-traveling waves exist at the load, then

$$V^-\left(t - \frac{\mathscr{L}}{u}\right) = -R_c I^-\left(t - \frac{\mathscr{L}}{u}\right) \tag{4.31}$$

Neither (4.30) nor (4.31) satisfies (4.29), but a combination of the two can satisfy it. However, (4.29) is also satisfied by (4.30) if $R_L = R_C$; in that case there is no backward-traveling wave and the line is said to be *matched* at the load. But the discontinuity in the line produced by the load resistor then results in a wave being reflected in the form of a backward-traveling wave.

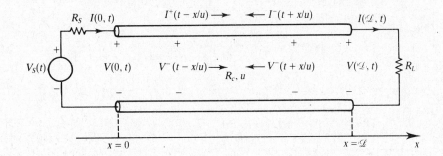

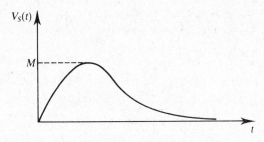

Fig. 4-6.

Fig. 4-7.

Reflection Coefficients

We may now define a *voltage reflection coefficient* at the load as the ratio of the amplitudes of the backward- and forward-traveling voltage waves at $x = \mathscr{L}$; that is,

$$\Gamma_{LV} = \frac{V^-(t + \mathscr{L}/u)}{V^+(t - \mathscr{L}/u)} = \Gamma_L \tag{4.32}$$

In terms of Γ_L and R_c, (4.22) and (4.28) give us, at the load,

$$V(\mathscr{L}, t) = V^+\left(t - \frac{\mathscr{L}}{u}\right)(1 + \Gamma_L) \tag{4.33}$$

$$I(\mathscr{L}, t) = \frac{V^+(t - \mathscr{L}/u)}{R_c}(1 - \Gamma_L) \tag{4.34}$$

Hence, a *current reflection coefficient* at the load may be defined as

$$\Gamma_{LI} \equiv \frac{I^-(t + \mathscr{L}/u)}{I^+(t - \mathscr{L}/u)} = -\Gamma_L \tag{4.35}$$

The current reflection coefficient is thus the negative of the voltage reflection coefficient. Solving (4.29) to (4.32) for R_L and Γ_L we obtain

$$R_L = R_c \frac{1 + \Gamma_L}{1 - \Gamma_L} \tag{4.36}$$

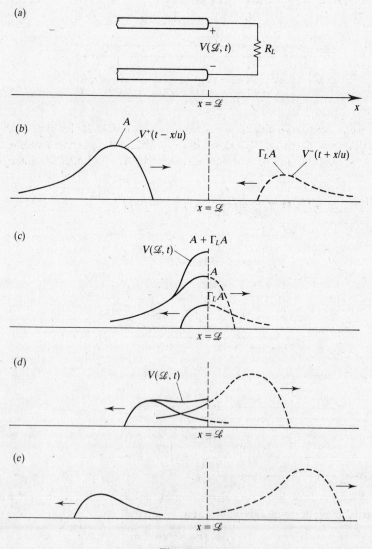

Fig. 4-8.

and
$$\Gamma_L = \frac{R_L - R_c}{R_L + R_c} \tag{4.37}$$

This reflection of waves is illustrated in Fig. 4-8. The reflection mechanism can be viewed as a mirror that produces, as the reflected wave V^- (shown dashed), a replica of V^+ that is "flipped around" and such that all points on the V^- waveform are the corresponding points of the V^+ waveform multiplied by Γ_L. Parts (b) through (e) of the figure show the waves at succeeding instants. At any time t, the total voltage at the load, $V(\mathscr{L}, t)$, is the sum of the individual waves present at the load at that time. This is illustrated in Fig. 4-8(b) and (c) for a forward-traveling wave of amplitude A.

Now let us consider the portion of the line at the source, $x = 0$, shown in Fig. 4-9(a). When the source is initially connected to the line, a forward-traveling wave is propagated down the line. A backward-traveling wave does not appear in the line until this initial forward-traveling wave has reached the load, which requires a time $T = \mathscr{L}/u$, since the load is presumed to have no source within it to produce a backward-traveling wave. That portion of the incident wave which is reflected at the load will require an additional time T to move from the load back to the source at $x = 0$. Therefore, during the interval $0 \le t < 2\mathscr{L}/u$, no backward-traveling wave will appear at $x = 0$, and the total voltage and current at $x = 0$ will be those due only to the forward-traveling waves V^+ and I^+; therefore,

$$V(0, t) = V^+\left(t - \frac{0}{u}\right) \tag{4.38}$$

$$\text{for } 0 \le t < \frac{2\mathscr{L}}{u}$$

$$I(0, t) = \frac{V^+(t - 0/u)}{R_c} \tag{4.39}$$

Since the ratio of the total voltage to the total current on the line is R_c for $0 \le t < 2\mathscr{L}/u$, the line appears to have an input resistance of R_c over this time interval, as shown in the circuit of Fig. 4-9(b). Thus, during this interval, the forward-traveling wave that is initially launched is related to $V_S(t)$ by

$$V(0, t) = \frac{R_c}{R_c + R_S} V_S(t) \qquad 0 \le t < \frac{2\mathscr{L}}{u} \tag{4.40}$$

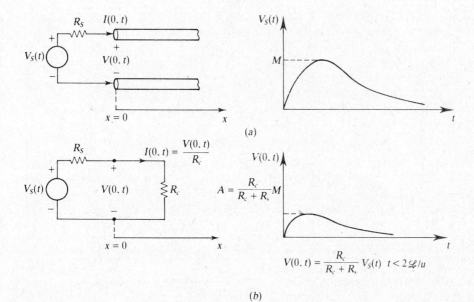

Fig. 4-9.

The initially launched wave is of the same shape as $V_S(t)$, but its points are reduced in magnitude from the corresponding points of $V_S(t)$ by the voltage-divider relationship $R_c/(R_c + R_S)$, as shown in the curve of Fig. 4-9(b). If M is the maximum amplitude of $V_S(t)$, then $A = R_c M/(R_c + R_S)$ is the maximum of $V(0, t)$.

This initially launched wave travels toward the load, requiring a time $T = \mathcal{L}/u$ for its leading edge to reach the load. When the leading edge does reach the load, a reflected wave is initiated, as shown in Fig. 4-8. This reflected wave requires a time $T = \mathcal{L}/u$ for its leading edge to reach the source. At the source, as at the load, this wave is reflected, and in parallel with (4.37) we can define a source voltage reflection coefficient

$$\Gamma_S = \frac{R_S - R_c}{R_S + R_c} \qquad (4.41)$$

as the ratio of the amplitude of the incoming wave (which was actually reflected at the load) to that of the reflected portion of this wave (which is sent back toward the load). A forward-traveling wave is therefore initiated at the source in the same fashion as a backward-traveling wave was initiated at the load. This forward-traveling wave has the same shape as the incoming backward-traveling wave, but with corresponding points reduced by Γ_S. This process of repeated reflection continues as re-reflection at the source and the load. At any time, the total voltage (or current) at any point on the line is the sum of the values for all the individual voltage waves (or current waves) existing on the line at that point and time.

Lattice Diagram

A convenient way of keeping track of these reflections is with a *lattice diagram* (Fig. 4-10). In the lattice diagram, the horizontal axis is labeled as distance down the line, and the vertical axis is labeled as time in increments of the total time required to transit the line in one direction: $\mathcal{L}/u$. Suppose the pulse shown in Fig. 4-10(b) is initially launched at time $t = 0$. Let us examine a point on this pulse having magnitude K at time t'. The point travels toward the load and is reflected, resulting in a corresponding point on the reflected waveform of magnitude $K\Gamma_L$. It is re-reflected at the source, then at the load, and so on. The lattice diagram shows this process in a convenient manner and allows us to obtain the value of the total line voltage $V(x, t)$ at any point on the line and any time. For example, at $t = t' + \mathcal{L}/u$, the total voltage at $x = \mathcal{L}$ is $K + K\Gamma_L = K(1 + \Gamma_L)$. At $x = \mathcal{L}/2$, a point midway down the line, at $t = t' + \frac{3}{2}\mathcal{L}/u$ the total line voltage is $K\Gamma_L$.

By following the movement of various points on the initially launched waveform [which differs from $V_S(t)$ only by $R_c/(R_c + R_S)$], we can sketch the total voltage at any point on the line. This can be done for the line current also, but then the initially launched wave is

$$I^+\left(t - \frac{0}{u}\right) = \frac{V_S(t)}{R_c + R_S} \qquad (4.42)$$

and we must replace Γ_S and Γ_L in the lattice diagram with the current reflection coefficients $-\Gamma_S$ and $-\Gamma_L$. However, the simplest method of sketching the line voltage and current is to visualize and sketch the individual forward- and backward-traveling waves and to combine all those present at an instant to produce the total line voltage and current distributions at that instant.

Solved Problems

4.1   Draw a phasor diagram showing the voltage-current relationships for a short transmission line represented by Fig. 4-1.

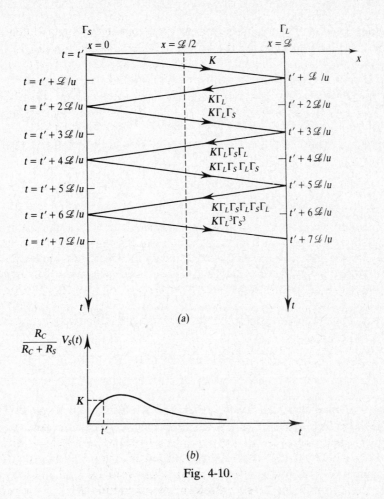

(a)

(b)

Fig. 4-10.

Choosing V_R as the reference phasor and arbitrarily choosing a power factor of 0.9, we draw the current I lagging V_R as shown in Fig. 4-11. For a short line, $I = I_S = I_R$ as shown. Also,

$$V_S = V_R + I(R + jX) \qquad \text{where } X = \omega L \tag{1}$$

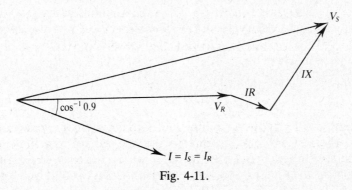

Fig. 4-11.

4.2 A 60-Hz short transmission line, having $R = 0.62$ ohms per phase and $L = 93.24$ millihenrys per phase, supplies a three-phase, wye-connected 100 MW load of 0.9 lagging power factor at 215 kV line-to-line voltage. Calculate the sending-end voltage per phase.

The line current $I\ (= I_S = I_R)$ is

$$I = \frac{100 \times 10^6}{\sqrt{3} \times 215 \times 10^3 \times 0.9} = 298.37 \text{ A}$$

and the per-phase voltage at the receiving end is

$$V_R = \frac{215 \times 10^3}{\sqrt{3}} = 124.13 \text{ kV}$$

The phasor diagram illustrating the operating conditions is that of Fig. 4-11, with $R = 0.62\ \Omega$ and $X = \omega L = 377 \times 93.24 \times 10^{-3} = 35.15\ \Omega$. Hence,

$$V_S = V_R + I(R + jX)$$
$$= 124.13 \times 10^3\underline{/0°} + (298.37\underline{/-25.8°})(0.62 + j35.15)$$
$$\approx 124.13 \times 10^3\underline{/0°} + (298.37\underline{/-25.8°})(35.15\underline{/90°})$$
$$= (128.69 + j9.44)\text{ kV} \approx 129.04\underline{/4.2°}\text{ kV}$$

4.3 Determine the voltage regulation and efficiency of transmission of the transmission line of Problem 4.2.

Since $V_{R(\text{no load})} = V_S$ here, (4.1) gives us

$$\text{Percent voltage regulation} = \frac{|V_S| - |V_R|}{|V_R|} \times 100 = \frac{129.04 - 124.13}{124.13} \times 100$$
$$= 3.955 \text{ percent}$$

To calculate the efficiency, we first determine the loss in the line, which is

$$\text{Line loss} = 3I^2R = 3 \times 298.37^2 \times 0.62 = 0.166 \text{ MW}$$

The power received at the load is given to be 100 MW, so the power sent is $100 + 0.166 = 100.166$ MW. Then, by (4.2),

$$\text{Efficiency} = \frac{100}{100.166} = 99.83 \text{ percent}$$

4.4 A 10-km-long, single-phase short transmission line has $0.5\underline{/60°}\ \Omega/\text{km}$ impedance. The line supplies a 316.8-kW load at 0.8 power factor lagging. What is the voltage regulation if the receiving-end voltage is 3.3 kV?

To find the voltage regulation, we must determine $V_S = V_{R(\text{no load})}$, for which we use (1) of Problem 4.1:

$$\cos^{-1}0.8 = 36.87°$$

and $$Z = (0.5\underline{/60°})(10) = 5\underline{/60°}\ \Omega$$

Then $$I = \frac{316.8 \times 10^3}{3.3 \times 10^3 \times 0.8}\underline{/-36.87°} = 120\underline{/-36.87°}\text{ A}$$

and $$IZ = (5\underline{/60°})(120\underline{/-36.87°}) = (551.77 + j235.69)\text{ V}$$

Now $$V_S = (3300 + j0) + (551.77 + j235.69) = (3851.77 + j235.69)\text{ V}$$

so $$|V_S| = 3858.97 \text{ V}$$

Hence, by (4.1),

$$\text{Percent voltage regulation} = \frac{3858.97 - 3300}{3300} \times 100 = 16.94 \text{ percent}$$

4.5 Calculate the sending-end power for the line of Problem 4.4 using (a) a loss calculation and (b) the sending-end voltage and power factor.

(a) The real part of Z is the resistance. Thus, since

$$Z = 5\underline{/60°} = (2.5 + j4.33)\,\Omega$$

we have   Line loss $= I^2R = 120^2 \times 2.5 = 36\,\text{kW}$

and   Sending-end power $= 316.8 + 36 = 352.8\,\text{kW}$

(b) From Problem 4.4, $V_S = 3858.97\underline{/3.5°}\,\text{V}$. As is shown in Fig. 4-11, the angle between V_S and I_S is the sum of this angle and the angle between V_R and I_S, or $3.5° + 36.87° = 40.37°$. Then

$$\text{Sending-end power} = \frac{3858.97 \times 120 \cos 40.37°}{1000} = 352.8\,\text{kW}$$

4.6   The per-phase impedance of a short transmission line is $(0.3 + j0.4)\,\Omega$. The sending-end line-to-line voltage is 3300 V, and the load at the receiving end is 300 kilowatts per phase at 0.8 power factor lagging. Calculate (a) the receiving-end voltage and (b) the line current.

(a) On a per-phase basis,

$$V_S = \frac{3300}{\sqrt{3}} = 1905.25\,\text{V} \tag{1}$$

and $$I = \frac{300 \times 10^3}{(0.8)V_R} = \frac{3.75 \times 10^5}{V_R}\,\text{A} \tag{2}$$

From Fig. 4-11, redrawn with I as reference phasor in Fig. 4-12, we determine that

$$V_S^2 = (V_R \cos \phi_R + RI)^2 + (V_R \sin \phi_R + XI)^2 \tag{3}$$

Substituting (1), (2), and other known values into (3) yields

$$1905.25^2 = \left(0.8V_R + \frac{0.3 \times 3.75 \times 10^5}{V_R}\right)^2 + \left(0.6V_R + \frac{0.4 \times 3.75 \times 10^5}{V_R}\right)^2$$

from which we find that $V_R = 1805\,\text{V}$.

(b) From (2), we have

$$I = \frac{3.75 \times 10^5}{1905} = 207.75\,\text{A}$$

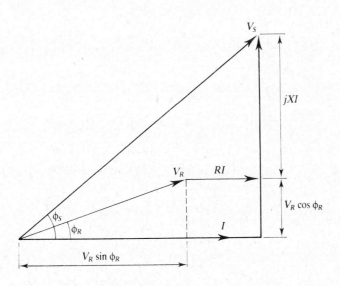

Fig. 4-12.

4.7 For the line of Problem 4.6, (a) calculate the sending-end power factor; (b) calculate the power loss per phase by determining the sending-end power; and (c) verify your result in part (b) by calculating the power loss directly.

(a) From Fig. 4-12 and Problem 4.6,

$$V_R \cos \phi_R + RI = (1805)(0.8) + (0.3)(207.75) = 1506.33 \, V$$

and $V_S = 1905.26 \, V$. Hence

$$\cos \phi_S = \frac{1506.33}{1905.25} = 0.79$$

and the sending end has 0.79 power factor lagging.

(b) The sending-end power is

$$P_S = V_S I \cos \phi_S = 1905.25 \times 207.75 \times 0.79 = 312.94 \, kW$$

The power loss per phase is then

$$P_S - P_R = 312.94 - 300 = 12.94 \, kW$$

(c) By direct calculation, the per-phase power loss in the line is

$$I^2 R = (207.75)^2(0.3) = 12.948 \, kW$$

4.8 For a given short transmission line of impedance $(R + jX)$ ohms per phase, the sending-end and receiving-end voltages, V_S and V_R respectively, are fixed. Find the maximum power that can be transmitted over the line.

From (3) of Problem 4.6, we have

$$V_S^2 = V_R^2 + 2IV_R(R \cos \phi_R + X \sin \phi_R) + I^2(R^2 + X^2) \tag{1}$$

Now, since

$$P = V_R I \cos \phi_R \quad \text{and} \quad Q = V_R I \sin \phi_R$$

we may rewrite (1) as

$$-V_S^2 + V_R^2 + 2PR + 2QX + \frac{1}{V_R^2}(P^2 + Q^2)(R^2 + X^2) = 0 \tag{2}$$

In (2) only P and Q vary. Thus, at maximum power we will have $dP/dQ = 0$. Differentiating (2) and rearranging yield

$$\frac{dP}{dQ} = -\frac{2X + 2QK}{2R + 2PK} \quad \text{where } K = \frac{R^2 + X^2}{V_R^2}$$

from which we find that, for $dP/dQ = 0$,

$$Q = -\frac{V_R^2 X}{R^2 + X^2} \tag{3}$$

Substituting (3) into (2) yields, after some algebraic simplification,

$$P_{max} = \frac{V_R^2}{Z^2}\left(\frac{ZV_S}{V_R} - R\right) \tag{4}$$

where $Z = \sqrt{R^2 + X^2}$.

4.9 What is the maximum power that can be transmitted over a three-phase short transmission line having a per-phase impedance of $(0.3 + j0.4) \, \Omega$ if the receiving-end voltage is 6351 volts per phase and the voltage regulation is not to exceed 5 percent?

On a per-phase basis,

$$V_R = 6351 \text{ V}$$
$$V_S = (1 + 0.05)(6351) = 6668.6 \text{ V}$$
$$Z = \sqrt{(0.3)^2 + (0.4)^2} = 0.5 \ \Omega$$

Then, from (4) of Problem 4.8,

$$P_{max} = \left(\frac{6351}{0.5}\right)^2\left(\frac{0.5 \times 6668.6}{6351} - 0.3\right) = 36.3 \text{ MW/phase}$$

and the maximum total power that can be transmitted is $3 \times 36.3 = 108.9$ MW.

4.10 Calculate (a) the receiving-end power factor and (b) the total line loss for the transmission line of Problem 4.9 while it is supplying maximum power.

(a) From (3) of Problem 4.8, the reactive power received per phase is

$$Q = -\frac{6351^2 \times 0.4}{0.5^2} = -64.53 \text{ Mvar}$$

Thus,
$$\tan \phi_R = \frac{64.53}{36.3} = 1.78$$

and $\cos \phi_R = 0.49$. Hence the receiving-end power factor is 0.49 lagging.
(b) From Problem 4.9, at maximum power

$$I = \frac{36.3 \times 10^6}{6351 \times 0.49} = 11,664 \text{ A/phase}$$

and
$$\text{Total line loss} = 3I^2R = 3(11,664)^2(0.3) = 122.46 \text{ MW}$$

4.11 The per-phase parameters for a 60-Hz, 200-km-long transmission line are $R = 2.07 \ \Omega$, $L = 310.8$ mH, and $C = 1.4774 \ \mu$F. The line supplies a 100-MW, wye-connected load at 215 kV (line-to-line) and 0.9 power factor lagging. Calculate the sending-end voltage, using the nominal-Π circuit representation.

To use the nominal-Π circuit, we first express V_R and I_R per phase as follows:

$$V_R = \frac{215 \times 10^3}{\sqrt{3}} = 124.13 \text{ kV}$$

$$I_R = \frac{100 \times 10^6}{\sqrt{3} \times 215 \times 10^3 \times 0.9} = 298.37\underline{/-25.8°} \text{ A}$$

Using the nomenclature of Fig. 4-2, we have

$$I_{CR} = \frac{V_R}{X_{C/2}} = \frac{124.13 \times 10^3\underline{/0°}}{1/(377 \times 0.5 \times 1.4774 \times 10^{-6})\underline{/90°}} = 34.57\underline{/90°} \text{ A}$$

$$I = I_R + I_{CR} = 298.37\underline{/-25.8°} + 34.57\underline{/90°} = 285\underline{/-29.5°} \text{ A}$$

$$R + jX_L = 2.07 + j377 \times 0.3108 \approx 117.19\underline{/88.98°} \ \Omega$$

$$I(R + jX_L) = 285\underline{/-19.5°} \times 117.19\underline{/88.98°} = 33.4\underline{/69.48°} \text{ kV}$$

Hence
$$V_S = V_R + I(R + jX_L) = 124.13\underline{/0°} + 33.4\underline{/69.48°}$$
$$= 139.39\underline{/12.97°} \text{ kV/phase}$$

4.12 Repeat Problem 4.11, using the nominal-T circuit representation for the transmission line.

Using the nomenclature of Fig. 4-3, we have, with V_R and I_R as computed in Problem 4.11,

$$V_C = V_R + \tfrac{1}{2}I_R(R + jX_L) = 124.13\underline{/0°} + \frac{10^{-3}}{2} \times 298.37\underline{/-25.8°} \times 117.19\underline{/88.98°}$$

$$= 132.92\underline{/6.74°} = (132 + j15.6)\ \text{kV}$$

Then
$$I_C = \frac{V_C}{X_C} = \frac{132.92 \times 10^3\underline{/6.74°}}{1/[(377 \times 1.4774 \times 10^{-6})/\underline{90°}]} = 74\underline{/96.74°}\ \text{A}$$

$$I_S = I_R + I_C = 298.37\underline{/-25.8°} + 74\underline{/96.74°} = 226.0\underline{/-12.2°}\ \text{A}$$

$$V_S = V_C + \tfrac{1}{2}I_S(R + jX_L) = 132.92\underline{/6.74°} + \frac{10^{-3}}{2} \times 266.0\underline{/-12.2°} \times 117.19\underline{/88.98°}$$

$$= 139.0\underline{/12.78°}\ \text{kV/phase}$$

4.13 The per-unit-length parameters of a 215-kV, 400-km, 60-Hz, three-phase long transmission line are $y = j3.2 \times 10^{-6}$ S/km and $z = (0.1 + j0.5)\ \Omega$/km. The line supplies a 150-MW load at unity power factor. Determine (*a*) the voltage regulation, (*b*) the sending-end power, and (*c*) the efficiency of transmission.

We will need V_S and I_S. Because this is a long line, with parameters assumed to be distributed along the line, we find the sending-end voltage and current as follows: we have

$$z = 0.1 + j0.5 = 0.51\underline{/78.7°}$$

and
$$y = j3.2 \times 10^{-6} = 3.2 \times 10^{-6}\underline{/90°}$$

so
$$\gamma\mathscr{L} = \mathscr{L}\sqrt{zy} = 400\sqrt{0.51 \times 3.2 \times 10^{-6}}\underline{/\tfrac{1}{2}(90 + 78.7)°} = 0.51\underline{/84.35°}$$

$$= 0.05 + j0.5 = \alpha\mathscr{L} + j\beta\mathscr{L}\ \text{rad}$$

In addition,

$$Z_C = \sqrt{\frac{z}{y}} = \sqrt{\frac{0.51}{3.2 \times 10^{-6}}}\ \underline{/\tfrac{1}{2}(78.7 - 90)°} = 399.2\underline{/-5.65°}\ \Omega$$

$$V_R = \frac{215 \times 10^3}{\sqrt{3}} = 124.13\underline{/0°}\ \text{kV}$$

and
$$I_R = \frac{150 \times 10^6}{\sqrt{3} \times 215 \times 10^3} = 402.8\underline{/0°}\ \text{A}$$

Now, from (*4.10*), we have

$$\cosh \gamma\mathscr{L} = \cosh 0.05 \cos 0.5 + j \sinh 0.05 \sin 0.5 = 0.877 + j0.024$$

$$= 0.877\underline{/1.57°}$$

$$\sinh \gamma\mathscr{L} = \sinh 0.05 \cos 0.5 + j \cosh 0.05 \sin 0.5 = 0.044 + j0.479$$

$$= 0.48\underline{/84.75°}$$

Finally, from (*4.8*) and (*4.9*) respectively, we obtain

$$V_S = (124.13\underline{/0°} \times 0.877\underline{/1.57°} + 402.8 \times 10^{-3}\underline{/0°} \times 399.2\underline{/-5.65°} \times 0.48\underline{/84.75°})\ \text{kV}$$

$$= 146.4\underline{/32.55°}\ \text{kV}$$

and
$$I_S = \left(\frac{124.13 \times 10^3\underline{/0°}}{399.2\underline{/-5.65°}} \times 0.48\underline{/84.75°} + 402.8\underline{/0°} \times 0.877\underline{/1.57°}\right)\text{A}$$

$$= 386.52\underline{/24.28°}\ \text{A}$$

(a) To use (4.1), we need the receiving-end voltage at no load, which we obtain from (4.8) with $I_R = 0$:

$$V_{R(\text{no load})} = \frac{|V_S|}{|\cosh \gamma \mathscr{L}|} = \frac{146.4}{0.877} = 166.93 \text{ kV}$$

Then   Percent voltage regulation $= \dfrac{166.93 - 124.13}{124.13} \times 100 = 34.48$ percent

(b) The sending-end power is

$$\text{Power sent} = 3V_S I_S \cos \phi$$

$$= 3 \times 146.4 \times 10^3 \times 386.5 \cos (32.55° - 24.28°)$$

$$= 167.98 \text{ MW}$$

(c) From (4.2), the efficiency of the transmission is

$$\text{Efficiency of transmission} = \frac{\text{power received}}{\text{power sent}} \times 100 = \frac{150}{167.98} = 89.3 \text{ percent}$$

4.14 Figure 4-13(a) is the phasor diagram corresponding to (4-17). By shifting the origin from O' to O, we turn Fig. 4-13(a) into a power diagram, as shown in detail in Fig. 4-13(b). For a given fixed value of $|V_R|$ and a set of values for $|V_S|$, draw the loci of the point A.

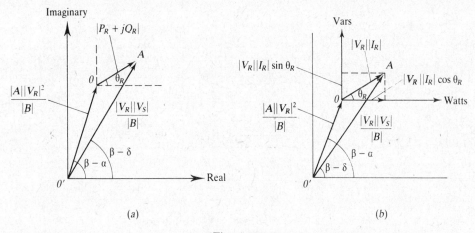

(a)   (b)

Fig. 4-13.

Because $O'A = |V_R| |V_S| / |B|$ for a given load and a given value of $|V_R|$, the loci of point A will be a set of circles (of radii $O'A$), one for each of the set of values of $|V_S|$. Portions of two such circles are given in Fig. 4-14; the circles are sometimes called *receiving-end circles*.

4.15 From the result of Problem 4.14 (Fig. 4-14), for a given load with a lagging power-factor angle θ_R, determine the amount of reactive power that must be supplied at the receiving end to maintain a constant receiving-end voltage, if the sending-end voltage decreases from $|V_{S1}|$ to $|V_{S2}|$.

Line OA in Fig. 4-14 is the load line whose intersection with the power circle determines the operating point. Thus, for a load having a lagging power-factor angle θ_R, A and C are, respectively, the operating points for sending-end voltages $|V_{S1}|$ and $|V_{S2}|$. These operating points determine the real and reactive power received for the two sending-end voltages.

The reactive power that must be supplied at the receiving end in order to maintain constant $|V_R|$

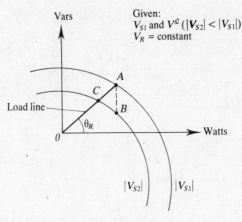

Fig. 4-14.

when the sending-end voltage decreases from $|V_{S1}|$ to $|V_{S2}|$ is given by the length AB, which is parallel to the reactive-power axis. (It may be supplied via capacitors in parallel with the load.)

4.16 Determine the $ABCD$ constants for the nominal-T circuit of a transmission line for which $R = 10\,\Omega$, $X = 20\,\Omega$, and $Y = 400\,\mu S$ for each phase.

From Table 4-1,

$$A = D = (1 + \tfrac{1}{2}YZ) = 1 + j\frac{4 \times 10^{-4}}{2}(10 + j20) = 0.996\underline{/0.115°}$$

$$B = Z(1 + \tfrac{1}{4}YZ) = (10 + j20)[1 + \tfrac{1}{4}(j4 \times 10^{-4})(10 + j20)]$$

$$= 22.25\underline{/63.45°}\,\Omega$$

$$C = Y = j4 \times 10^{-4} = 4 \times 10^{-4}\underline{/90°}\,S$$

4.17 Determine the $ABCD$ constants for the line of Problem 4.11. Then rework the problem, treating the line as a two-port network.

From Problem 4.11,

$$Z = 117.19\underline{/88.98°}\,\Omega$$

and

$$Y = 5.57 \times 10^{-4}\underline{/90°}\,S$$

Then, from Table 4-1,

$$A = D = 1 + \tfrac{1}{2}YZ = 1 + \tfrac{1}{2}(117.19\underline{/88.98°})(5.57 \times 10^{-4}\underline{/90°}) = 0.967\underline{/0.034°}$$

$$B = Z = 117.19\underline{/99.09°}\,\Omega$$

$$C = Y(1 + \tfrac{1}{4}YZ) = 5.57 \times 10^{-4}\underline{/90°}\,[1 + \tfrac{1}{4}(5.57 \times 10^{-4}\underline{/90°})(117.19\underline{/88.98°})]$$

$$= 5.54 \times 10^{-4}\underline{/-89.96°}\,S$$

Finally, from (*4.11*) and Problem 4.11, the sending-end voltage is

$$V_S = AV_R + BI_R = (0.967\underline{/0.034°})(124.13\underline{/0°}) + 10^{-3}(117.19\underline{/88.98°})(298.37\underline{/-25.8°})$$

$$= 120.03\underline{/0.034°} + 34.96\underline{/63.18°} = 139.32\,kV/phase$$

4.18 At time $t = 0$, a 30-V battery with zero source resistance is connected to the transmission line shown in Fig. 4-15(*a*). Sketch the distribution of voltage along the line for several instants of time.

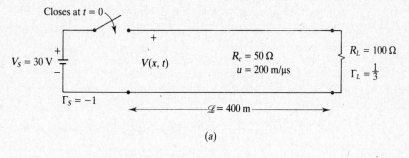

(a)

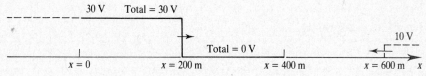

(b) $t = 1$ μs

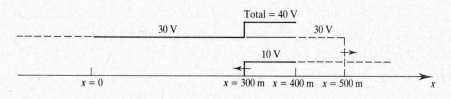

(c) $t = 2.5$ μs

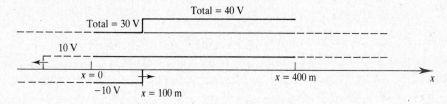

(d) $t = 4.5$ μs

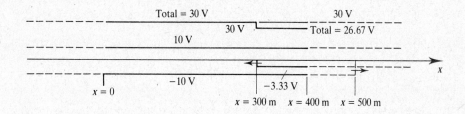

(e) $t = 6.5$ μs

Fig. 4-15.

The load and source voltage reflection coefficients are

$$\Gamma_L = \frac{R_L - R_c}{R_L + R_c} = \tfrac{1}{3}$$

$$\Gamma_S = \frac{R_S - R_c}{R_S + R_c} = -1$$

and the time required to transit the line in one direction is $\mathscr{L}/u = 2\,\mu s$. At $t = 0$, a 30-V pulse is sent down the line; at points along the transmission line, the voltage is zero prior to the arrival of the pulse and 30 V after the pulse has passed. Figure 4-15(b) shows the voltage at $t = 1\,\mu s$. At, $t = 2.5\,\mu s$, the pulse has already arrived at the load, and a backward-traveling pulse of magnitude $30\Gamma_L = 10\,V$ has been sent back toward the source. Figure 4-15(c) shows the situation at that time.

When this 10-V reflected pulse arrives at the source, at $t = 4\,\mu s$, a pulse of magnitude $\Gamma_S \times 10\,V = -10\,V$ is sent back toward the load. Figure 4-15(d) shows the situation 0.5 μs later, at $t = 4.5\,\mu s$.

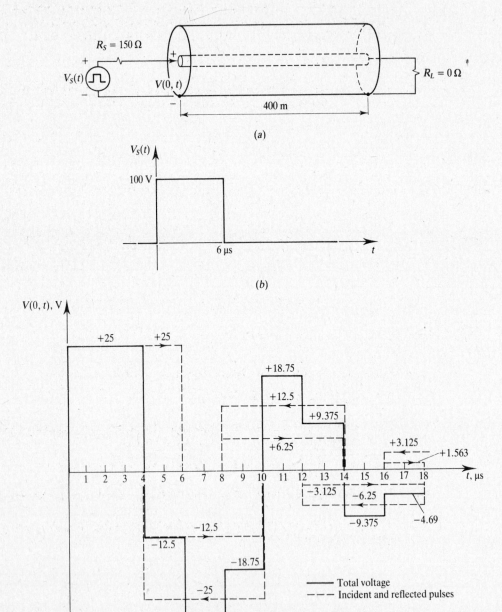

Fig. 4-16.

The -10-V pulse travels to the load, reaching it at $t = 6\,\mu s$, at which time a reflected pulse of magnitude $\Gamma_L \times -10 = \Gamma_L \times \Gamma_S \times \Gamma_L \times 30 = -3.33\,V$ is sent back toward the source. Figure 4-15(e) shows the situation $6.5\,\mu s$ later. At each point on the line, at any time, the total line voltage is the sum of the voltage waves present at that point and at that time.

4.19   The 400-m length of cable in Fig. 4-16(a) terminates in a short circuit ($R_L = 0$) and is driven by a pulse source having an internal resistance of $150\,\Omega$ (that is, $R_S = 150\,\Omega$). The source produces a pulse of magnitude 100 V and duration $6\,\mu s$, as shown in Fig. 4-16(b). Sketch the voltage $V(0, t)$ at the input to the line for the first $18\,\mu s$. The cable parameters are $C = 100\,pF/m$ and $L = 0.25\,\mu H/m$.

The characteristic resistance is

$$R_c = \sqrt{\frac{L}{C}} = \sqrt{\frac{0.25 \times 10^{-6}}{100 \times 10^{-12}}} = 50\,\Omega$$

and the velocity of propagation is

$$u = \frac{1}{\sqrt{LC}} = 200 \times 10^6\,m/s = 200\,m/\mu s$$

Consequently, the time required for the pulse to move from one end of the cable to the other is $2\,\mu s$.
The voltage reflection coefficient at the load is

$$\Gamma_L = \frac{R_L - R_c}{R_L + R_c} = -1$$

and that at the source is

$$\Gamma_S = \frac{R_S - R_c}{R_S + R_c} = \frac{1}{2}$$

The source initially sees an input resistance to the line of $R_c = 50\,\Omega$. Thus, the initially launched voltage wave is a pulse that has a 6-μs duration and a magnitude of

$$\frac{R_c}{R_c + R_S}\,V_S = 25\,V$$

This pulse reaches the load in $2\,\mu s$, at which time and place a pulse with magnitude $\Gamma_L \times 25 = -25\,V$ is reflected; this reflected pulse is re-reflected $2\,\mu s$ later at the source, producing a pulse of magnitude $\Gamma_S \times -25 = -12.5\,V$ traveling back toward the load; and so forth. The dashed lines in Fig. 4.16(c) show the contributions of the various pulses to $V(0, t)$ as a function of time. The arrows indicate direction: $\rightarrow$ denotes a forward-traveling pulse, and $\leftarrow$ a backward-traveling pulse. The total voltage at the source end, plotted as a solid line, is the sum of all the voltages present at $x = 0$ at any time.

4.20   For the transmission line and source voltage of Problem 4.18, sketch the voltage $V(\mathscr{L}, t)$ at the load and the input current $I(0, t)$ as a function of time for the first $16\,\mu s$. Figure 4-17(a) shows the circuit and source voltage waveform.

At $t = 0$, a 30-V pulse is sent out by the source. The leading edge of this pulse arrives at the load at $t = 2\,\mu s$. At this time, a pulse of magnitude $\Gamma_L \times 30 = 10\,V$ is sent back toward the source. This 10-V pulse arrives at the source at $t = 4\,\mu s$, at which time a pulse of magnitude $\Gamma_S\Gamma_L \times 30 = -10\,V$ is reflected back to the load. This pulse arrives at the load at $t = 6\,\mu s$, at which time a pulse of magnitude $\Gamma_L\Gamma_S\Gamma_L \times 30 = -3.33\,V$ is sent back toward the source. The contributions of these waves at $x = \mathscr{L}$ are shown in Fig. 4-17(b) by dashed lines, and the total voltage there is shown by the solid line. Note that the load voltage oscillates about 30 V but asymptotically approaches 30 V.

To sketch the input current $I(0, t)$, we sketch the forward- and backward-traveling waves directly and add them to obtain the total input current, as is done in Fig. 4-17(c). It would be *absurd* to sketch the input voltage $V(0, t)$ and then divide this by R_c to get $I(0, t)$, since the ratio of total voltage to total

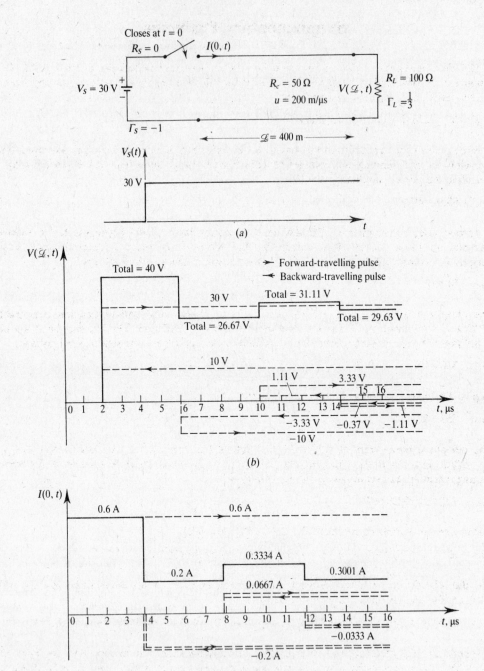

Fig. 4-17.

current on the line is *not* R_c except for $t < 2\mathscr{L}/u$. However, we could sketch $I(0, t)$ from a sketch of $V(0, t)$ by realizing that

$$I(0, t) = \frac{V_S(t) - V(0, t)}{R_S}$$

Thus, we could subtract $V(0, t)$ from $V_S(t)$ point by point, and divide the result by R_S to obtain $I(0, t)$. But $R_S = 0$ in this example, so we have no recourse other than to sketch $I(0, t)$ as in Fig. 4-17(c). Note there that $I(0, t)$ oscillates about the steady-state value of $30\,\text{V}/R_L = 0.3\,\text{A}$.

Supplementary Problems

4.21   A 138-kV, three-phase short transmission line has a per-phase impedance of $(2 + j4)\,\Omega$. If the line supplies a 25-MW load at 0.8 power factor lagging, calculate (a) the efficiency of transmission and (b) the sending-end voltage and power factor.

Ans.   (a) 98.78 percent; (b) 139.5 kV, 0.99

4.22   A three-phase short transmission line having a per-phase impedance of $(2 + j4)\,\Omega$ has equal line-to-line receiving-end and sending-end voltages of 115 kV while supplying a load at 0.8 power factor leading. Calculate the power supplied by the line.

Ans.   839.2 MW

4.23   A three-phase, wye-connected, 20-MW load of power factor 0.866 lagging is to be supplied by a transmission line at 138 kV. It is desired that the line losses not exceed 5 percent of the load. If the per-phase resistance of the line is $0.7\,\Omega$, what is the maximum length of the line?

Ans.   51 km

4.24   The per-phase constants of a 345-kV, three-phase, 150-km-long transmission line are resistance = $0.1\,\Omega/\text{km}$, inductance = 1.1 mH/km, and capacitance = $0.02\,\mu\text{F/km}$. The line supplies a 180-MW load at 0.9 power factor lagging. Using the nominal-Π circuit, determine the sending-end voltage.

Ans.   350.8 kV

4.25   Repeat Problem 4.24 using the nominal-T circuit.

Ans.   359.3 kV

4.26   The per-phase parameters of a 345-kV, 500-km, 60-Hz, three-phase transmission line are $y = j4 \times 10^{-6}\,\text{S/km}$ and $z = (0.08 + j0.6)\,\Omega/\text{km}$. If the line supplies a 200-MW load at 0.866 power factor lagging, calculate the sending-end voltage and power.

Ans.   372 kV; 240.8 MW

4.27   Determine the *ABCD* constants of the line of Problem 4.24.

Ans. $A = D = 0.965\underline{/0.5^\circ}$; $B = 64\underline{/76.4^\circ}\,\Omega$; $C = 0.982\underline{/0.25^\circ}$ S

4.28   List the *ABCD* constants and determine the sending-end voltage for the transmission line of Problem 4.26, considering the line as a two-port network.

Ans. $A = 0.7147\underline{/0^\circ}$, $B = 270.88\underline{/90^\circ}$, $C = 1.8 \times 10^{-3}\underline{/90^\circ}$, $D = 0.7147\underline{/0^\circ}$; $V_S = 372$ kV

4.29   The sending- and receiving-end voltages of a three-phase short transmission line are $V_S = 33$ kV and $V_R = 31.2$ kV, respectively. The per-phase line parameters are $R = 10\,\Omega$ and $X_L = 20\,\Omega$. Calculate the maximum power that can be transmitted by the line.

Ans.   26.57 MW

4.30   For a three-phase long transmission line, $Z_c = 406.4\underline{/-5.48^\circ}\,\Omega$, $V_R = 215\underline{/0^\circ}$ kV and $I_R = 335.7\underline{/0^\circ}$ A, all per phase. Calculate the sending-end line voltage.

Ans.   238.8 kV

4.31   How much power is transmitted over the line of Problem 4.30?

Ans.   137.5 MW

4.32　Evaluate the *ABCD* constants of the transmission line of Problem 4.30, and verify that $AD - BC = 1$.

　　　Ans.　$A = D = 0.89\underline{/1.34°}$, $B = 186.8\underline{/79.46°}$, $C = 0.00113\underline{/90.42°}$

4.33　A lossless transmission line has a 30-Ω characteristic resistance and terminates in a 90-Ω resistance. A 120-V dc source is applied at $t = 0$. Plot V_R versus time for this line, from $t = 0$ to $t = 5T$, where T is the time required for the voltage wave to travel the length of the line.

　　　Ans.　Fig. 4-18

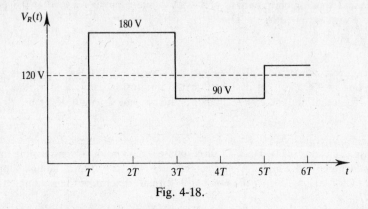

Fig. 4-18.

4.34　Rework Problem 4.33 by drawing a lattice diagram.

　　　Ans.　Fig. 4-19

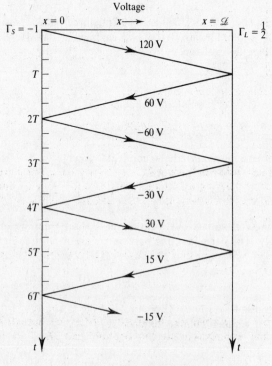

Fig. 4-19.

4.35　Draw the lattice diagram for the line of Problem 4.33, assuming the line now terminates in a 10-Ω resistance and the dc source has an internal resistance of $60\,\Omega$. Use it to plot $V(\mathcal{L}/3, t)$ versus t from $t = 0$ through $t = 4T$.

Ans. Fig. 4-20

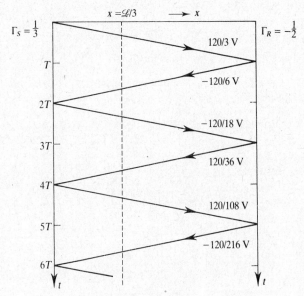

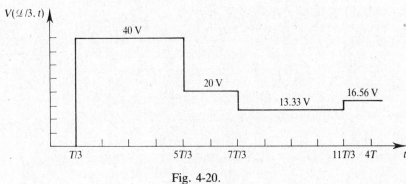

Fig. 4-20.

4.36 A 400-m-long lossless transmission line has a 100-Ω characteristic resistance and terminates in a 60-Ω resistance. A 400-V dc source having an internal resistance of $300\,\Omega$ is connected to the line at $t = 0$. Calculate the sending-end and receiving-end voltage reflection coefficients of the line.

Ans. $\frac{1}{2}$; $-\frac{1}{4}$

4.37 If the parameters of the line of Problem 4.36 are such that the velocity of wave propagation along the line is $400\,\text{m}/\mu\text{s}$, sketch $V_S(t)$ for $0 \le t \le 10\,\mu\text{s}$.

Ans. Fig. 4-21

4.38 Determine the sending-end and receiving-end current reflection coefficients of the line of Problem 4.36, and calculate the initial line current from the data of Problem 4.37.

Ans. $-\frac{1}{2}$; $\frac{1}{4}$; 1 A

4.39 From the data of Problems 4.36 and 4.37, sketch $V_R(t)$ for $0 \le t \le 10\,\mu\text{s}$.

Ans. Fig. 4-22

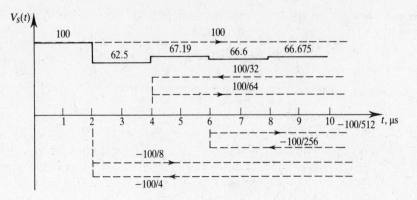

$$V_S(t)_{ss} = \frac{60}{300 + 60}\ 400 = 66.667\ \text{V}$$

Fig. 4-21.

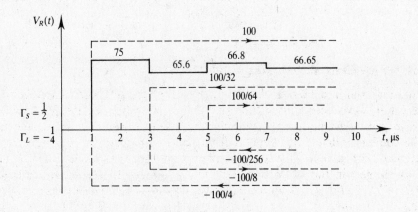

$$V_R(t)_{ss} = \frac{60}{300 + 60}\ 400 = 66.667\ \text{V} = V_S(t)_{ss}$$

Fig. 4-22.

4.40 Calculate the steady-state current and voltage at the receiving end of the line of Problems 4.36 and 4.37. Verify that the results are consistent with the sketch obtained in Problem 4.39.

Ans.   1.11 A; 66.666 V

4.41 Suppose the line of Problems 4.36 and 4.37 is short-circuited at the receiving end. Calculate the voltage and current reflection coefficients and the steady-state receiving-end voltage and current.

Ans.   Voltage reflection coefficients = $\frac{1}{2}$, -1; current reflection coefficients = $-\frac{1}{2}$, 1; steady-state $V_R = 0$; steady-state $I_R = 1.333$ A

4.42 Suppose the line of Problems 4.36 and 4.37 is open-circuited at the receiving end. Determine the voltage reflection coefficients and the steady-state receiving-end voltage and current.

Ans.   Voltage reflection coefficients = $\frac{1}{2}$, 1; steady-state $V_R = 400$ V; steady-state $I_R = 0$

Chapter 5

Underground Cables

Unlike overhead transmission lines, underground cables must have adequate electrical insulation to protect the conductor from contacting the ground or the cable's external shield. Furthermore, protection from mechanical, chemical, and other hazardous effects must be provided, to ensure satisfactory and reliable operation. The basic components of an underground cable are (1) the conductor, which is made of stranded copper or aluminum; (2) the insulation around the conductor, which may be some form of rubber such as vulcanized or butyl rubber, or a special-purpose synthetic such as polyvinyl chloride (PVC), or oil-impregnated paper; and (3) the external protective covering, which is often a lead alloy sheath applied over the insulated cable.

A three-phase underground cable has three conductors within the sheath. Because the conductors of a three-conductor cable are much closer to each other than those of an overhead line, and because the conductors are immersed in dielectrics, the capacitive reactance of a cable is smaller than that of a comparable transmission line. Thus, the use of a T or Π circuit representation for a cable of even short length may be necessary for analysis.

5.1 ELECTRIC STRESS IN A SINGLE-CORE CABLE

The insulating material in a cable constitutes a dielectric and has a certain dielectric strength. If this dielectric strength is exceeded during the operation of the cable, the insulation will break down. Hence the cable must be designed so that the electric field strength, or maximum electric stress, at the surface of the conductor does not exceed that required to break down the insulation. However, if the cable is designed with a relatively low line-to-ground voltage gradient, the overall size of the cable becomes too large; on the other hand, if the voltage gradient is made large so as to allow reduction of the cable diameter, then the dielectric loss may become too large and may lead to excessive heating of the cable.

It has been found that the optimal ratio of the radius of the cable to the radius of the conductor is given by

$$\frac{R_1}{R_2} = e = 2.718 \qquad (5.1)$$

where R_1 and R_2 are defined in Fig. 5-1. Smaller ratios will result in unstable cable operation, in that the dielectric will tend to break down. Any ratio exceeding 2.718 will result in satisfactory cable operation. For economic reasons, however, it is best to maintain R_1/R_2 close to 2.718.

5.2 GRADING OF CABLES

The cable in Fig. 5-1 is filled with a single layer of a single dielectric, but many cables contain several layers of dielectric. The dielectric materials in such a cable are chosen and distributed so as to minimize the difference between the maximum and minimum electric field strengths in the cable. This process is known as *grading,* and two types are commonly used—capacitance grading and intersheath grading.

In *capacitance grading,* two or more layers of different dielectrics are used to insulate a cable. Two such layers are shown in Fig. 5-2, for the sake of illustration. The permittivities of these layers are so chosen that the maximum field strength is the same in both regions. The corresponding variation of the electric field E with radius r is shown in Fig. 5-3. For equal maximum field

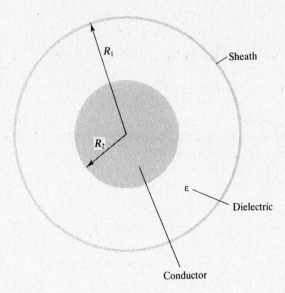

Fig. 5-1.

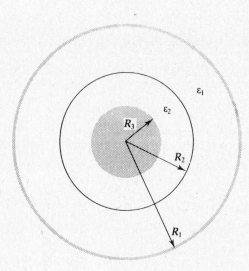

Fig. 5-2.

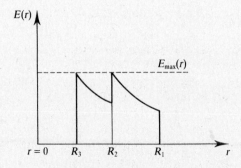

Fig. 5-3.

strengths, we must have (see Problem 5.4)

$$\epsilon_1 R_2 = \epsilon_2 R_3 \tag{5.2}$$

In that case, if E_{max} is the maximum allowable electric field, the operating voltage V of the cable is

$$V = E_{max}\left(R_3 \ln \frac{R_2}{R_3} + R_2 \ln \frac{R_1}{R_2}\right) \tag{5.3}$$

In *intersheath grading*, the cable contains several layers of a single dielectric material, separated by coaxial metallic sheaths that are inserted into the dielectric and maintained at predetermined voltages. A cable with one such intersheath is shown in cross section in Fig. 5-4. Let the radii R_1, R_2, and R_3 be such that

$$\frac{R_1}{R_2} = \frac{R_2}{R_3} = a \tag{5.4}$$

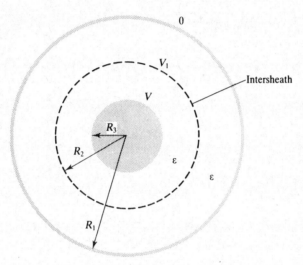

Fig. 5-4.

If the intersheath is kept at voltage V_1, then at the surface of the conductor we have

$$E_{3max} = \frac{V - V_1}{R_3 \ln (R_2/R_3)} = \frac{V - V_1}{R_3 \ln a} \tag{5.5}$$

At the surface of the intersheath, the maximum electric field is

$$E_{2max} = \frac{V_1}{R_2 \ln (R_1/R_2)} = \frac{V_1}{R_2 \ln a} \tag{5.6}$$

For the maximum electric fields to be the same at these two surfaces, we must have

$$V_1 = \left(\frac{a}{1 + a}\right)V \tag{5.7}$$

Now (5.5) and (5.7) yield

$$E_{3max} = \frac{V}{R_3 \ln a}\left(\frac{1}{1 + a}\right) \tag{5.8}$$

Without the intersheath, from Problem 5.1 and (5.4) we have

$$E_{max} = \frac{V}{R_3 \ln (R_1/R_3)} = \frac{V}{R_3 \ln \left(\dfrac{aR_2}{R_2/a}\right)} = \frac{V}{R_3 \ln a^2} = \frac{V}{2R_3 \ln a} \qquad (5.9)$$

Comparing (5.8) and (5.9), we find that

$$\frac{E_{3max} \text{ (with intersheath)}}{E_{max} \text{ (without intersheath)}} = \frac{2}{1 + a} \qquad (5.10)$$

5.3 CABLE CAPACITANCE

The capacitance per unit length of a single-conductor cable such as that in Fig. 5-1 is given by

$$C = \frac{Q}{V} = \frac{2\pi\epsilon}{\ln (R_1/R_2)} \quad \text{(in farads per meter)} \qquad (5.11)$$

In a three-conductor cable, the capacitances between pairs of conductors and between the conductors and the sheath are as shown in Fig. 5-5, where we assume the conductors are equilaterally spaced. To find the capacitance per phase, we change the delta-connected capacitances C_2 to their equivalent wye form as shown in Fig. 5-6(a), obtaining the capacitance combinations shown in Fig. 5-6(b) and (c). Figure 5-6(c) shows that the net capacitance per phase is $C_n = C_1 + 3C_2$.

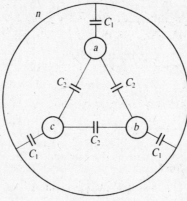

Fig. 5-5.

5.4 CABLE INDUCTANCE

The inductance per unit length of a single-conductor cable such as that in Fig. 5-1 is given by

$$L = \frac{\mu_0}{2\pi} \ln \frac{R_1}{R_2} \quad \text{(in henrys per meter)} \qquad (5.12)$$

Analytical expressions leading to the per-phase inductance of a three-conductor cable are extremely cumbersome and are not considered here.

5.5 DIELECTRIC LOSS AND HEATING

In an underground cable, heat is generated through I^2R losses in the conductor and sheath, and dielectric loss in the insulation. The dielectric loss in the insulation of the cable occurs due to leakage

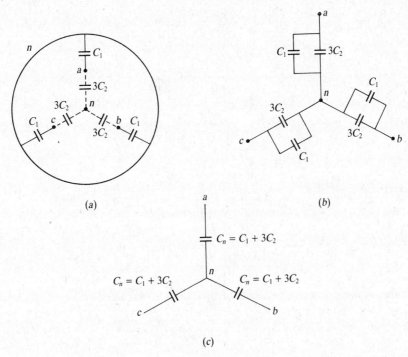

Fig. 5-6.

currents. In other words, the capacitance of the cable may be considered to be lossy, having a resistance R_i as shown in Fig. 5-7(a). The loss in R_i is

$$P = \frac{V^2}{R_i} \qquad (5.13)$$

In terms of the loss angle δ shown in Fig. 5-7(b), we have

$$\tan \delta = \frac{I_{Ri}}{I_C} = \frac{V/R_i}{\omega C V} \qquad (5.14)$$

Equations (5.13) and (5.14) now yield

$$P = \omega C V^2 \tan \delta \approx \omega C V^2 \delta \qquad (5.15)$$

if δ is small.

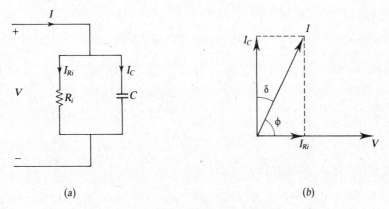

Fig. 5-7.

Solved Problems

5.1 Find the maximum and minimum electric field strengths in the cable shown in Fig. 5-1.

Let ρ_s be the surface charge density at the conductor surface. Then, for a unit length of the cable, within the dielectric and at a distance r (Fig. 5-8) from the center, we obtain from Gauss' law

$$2\pi R_2 \rho_s = D2\pi r = 2\pi\epsilon Er \tag{1}$$

where D is the electric flux density, E is the electric field strength and ϵ is the permittivity of the dielectric. We may rewrite (1) as

$$E = \frac{2\pi R_2 \rho_s}{2\pi\epsilon r} = \frac{Q}{2\pi\epsilon r} \tag{2}$$

where $Q = 2\pi R_2 \rho_s$ is the total charge per unit length at the conductor surface. From (2) we have, for the voltage V between the conductor and the sheath,

$$V = -\int_{R_1}^{R_2} E\, dr = \int_{R_2}^{R_1} \frac{Q}{2\pi\epsilon} \frac{dr}{r} = \frac{Q}{2\pi\epsilon} \ln \frac{R_1}{R_2} \tag{3}$$

Consequently, from (2) and (3) the electric field (which is entirely radial) is

$$E = \frac{V}{r \ln (R_1/R_2)} \tag{4}$$

The maximum and minimum values of E are then

$$E_{max} = \frac{V}{R_2 \ln (R_1/R_2)} \tag{5}$$

and

$$E_{min} = \frac{V}{R_1 \ln (R_1/R_2)} \tag{6}$$

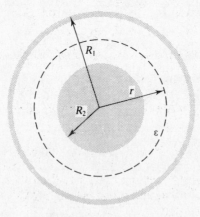

Fig. 5-8.

5.2 Derive (5.1).

To find the optimal ratio R_1/R_2, we fix R_1 and minimize E_{max} as given by (5) of Problem 5.1, as a function of R_2. For (5) to be a minimum, its denominator must be a maximum. For that, we must have

$$\frac{d}{dR_2} \left(R_2 \ln \frac{R_1}{R_2} \right) = \ln \frac{R_1}{R_2} - R_2 \frac{1}{R_1/R_2} \frac{R_1}{R_2^2} = 0 \tag{1}$$

From (*1*) it follows that $\ln (R_1/R_2) = 1$ and that

$$\frac{R_1}{R_2} = e^1 = 2.718 \tag{2}$$

5.3   A 13.2-kV single-conductor cable has an outside diameter of 10.0 cm. Determine the conductor radius and the electric field strength that must be withstood by the insulating material in the most economical (optimal-ratio) configuration.

From (*5.1*) or (*2*) of Problem 5.2, we obtain

$$\frac{R_1}{R_2} = \frac{10/2}{R_2} = 2.718$$

from which $R_2 = 1.84$ cm. Then, from (*5*) of Problem 5.1, we have

$$E_{max} = \frac{13.2 \times 1000}{1.84 \times 10^{-2} \ln (5/1.84)} = 717.62 \text{ kV/m}$$

5.4   For the cable shown in Fig. 5-2, obtain the condition under which the maximum values of the electric fields in the two regions are equal.

The electric fields in the two dielectrics are

$$E_1 = \frac{Q}{2\pi\epsilon_1 r} \qquad R_2 < r < R_1$$

and

$$E_2 = \frac{Q}{2\pi\epsilon_2 r} \qquad R_3 < r < R_2$$

The maximum values of these electric fields are at $r = R_2$ for E_1 and at $r = R_3$ for E_2. For the maximum values of the electric fields in the two regions to be same, we must have

$$E_{max} = \frac{Q}{2\pi\epsilon_1 R_2} = \frac{Q}{2\pi\epsilon_2 R_3}$$

or

$$\epsilon_1 R_2 = \epsilon_2 R_3 \tag{1}$$

Also, since $R_3 < R_2$, (*1*) implies that $\epsilon_2 > \epsilon_1$; that is, the dielectric closest to the conductor must have the highest permittivity.

5.5   Sketch the electric field distribution in the cable of Fig. 5-2 if (*1*) of Problem 5.4 is implemented.

The distribution is that of Fig. 5-3.

5.6   For the cable shown in Fig. 5-2, let $R_1 = 2.5$ cm, $R_3 = 0.92$ cm, and $R_2 = 1.75$ cm. Find the maximum electric field for an operating voltage of 13.2 kV, (*a*) with capacitance grading and (*b*) without capacitance grading.

(*a*) From (*5.3*) we obtain

$$13.2 \times 10^3 = E_{max}\left(0.92 \ln \frac{1.75}{0.92} + 1.75 \ln \frac{2.5}{1.75}\right) \times 10^{-2}$$

Hence, $E_{max} = 1085.8$ kV/m.
(*b*) Without capacitance grading, from (*5*) of Problem 5.1 we obtain (with R_2 replaced by $R_3 = 0.92$ cm)

$$E_{max} = \frac{13.2 \times 1000}{0.92 \ln (2.5/0.92) \times 10^{-2}} = 1435.3 \text{ kV/m}$$

5.7   A cable has intersheath grading that satisfies (*5.4*). The cable radii (see Fig. 5-4) are $R_3 = 0.92$ cm and $R_1 = 2.5$ cm. Determine the location of the intersheath, and calculate the ratio of maximum electric field strengths with and without intersheath.

The location of the intersheath (radius R_2 in Fig. 5-4) is given by (*5.4*), rewritten as

$$R_2^2 = R_1 R_3 = (2.5)(0.92)$$

from which $R_2 = 1.516$ cm.

From the known data, we have

$$a = \frac{R_1}{R_2} = \frac{2.5}{1.516} = 1.648$$

and, from (*5.10*),

$$\frac{E_{3\max} \text{ (with intersheath)}}{E_{\max} \text{ (without intersheath)}} = \frac{2}{1 + a} = \frac{2}{1 + 1.648} = 0.755$$

5.8   If the cable of Problem 5.7 is designed to operate nominally at 13.2 kV without any grading, determine the maximum voltage at which the cable may be operated with appropriate intersheath grading.

From Problem 5.7, $a = 1.648$, $R_2 = 1.516$ cm, and $R_3 = 0.92$ cm. Then, from (*5.8*), the maximum electric field with the intersheath grading is

$$E_{3\max} = \frac{V}{R_3 \ln a} \frac{1}{1 + a} = \frac{13.2}{0.92 \ln 1.648} \frac{1}{1 + 1.648}$$
$$= 16.09 \text{ kV/cm}$$

The maximum operating voltage is then given by (*5.3*), which is also valid in this case:

$$V = E_{3\max}\left(R_3 \ln \frac{R_2}{R_3} + R_2 \ln \frac{R_1}{R_2}\right) = E_{3\max}(R_3 + R_2) \ln a$$
$$= 16.09(0.92 + 1.516) \ln 1.648 = 19.58 \text{ kV}$$

5.9   Radially flowing leakage currents (dashed arrows in Fig. 5-9) are often present in underground cables. The leakage current is essentially limited by the insulation resistance of

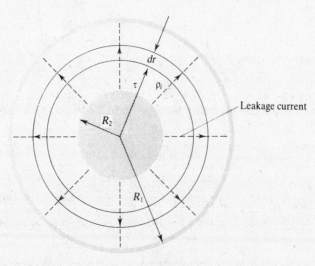

Fig. 5-9.

the cable. Derive an expression for the insulation resistance of a cable of length l meters and having a dielectric resistivity of ρ_i ohm-meters.

For an annular cylinder of thickness dr, as shown in Fig. 5-9, the elemental insulation resistance dR_i to leakage currents is

$$dR_i = \frac{\rho_i \, dr}{2\pi r l}$$

where l is the length of cable and ρ_i is the resistivity of the insulating material. The total insulation resistance is then

$$R_i = \frac{\rho_i}{2\pi l} \int_{R_2}^{R_1} \frac{dr}{r} = \frac{\rho_i}{2\pi l} \ln \frac{R_1}{R_2} \qquad (1)$$

5.10 In a test of a three-conductor cable, the three conductors are first bunched together, and the capacitance between the bunched conductors and sheath is found to be C_A. Then two of the conductors are bunched with the sheath, and the capacitance between these and the third conductor is found to be C_B. Determine C_1 and C_2 of Fig. 5-5.

From the first test we obtain the equation

$$C_A = 3C_1$$

The second test yields the equation

$$C_B = C_1 + 2C_2$$

Solving the two simultaneously yields

$$C_1 = \tfrac{1}{3}C_A$$
$$C_2 = \tfrac{1}{2}(C_B - \tfrac{1}{3}C_A)$$

5.11 The capacitances per kilometer of a three-wire cable are $0.90 \, \mu F$ between the three bunched conductors and the sheath, and $0.40 \, \mu F$ between one conductor and the other two connected to the sheath. Determine the line-to-ground capacitance of a 20-km length of this cable.

We have, in the terminology of Problem 5.10, $C_A = 0.90 \, \mu F/km$ and $C_B = 0.40 \, \mu F/km$. Then

$$C_1 = \tfrac{1}{3}C_A = 0.3 \, \mu F/km$$

and $$C_2 = \tfrac{1}{2}(0.4 - 0.3) = 0.05 \, \mu F/km$$

Now, from Fig. 5-6(c),

$$C_n = C_1 + 3C_2 = 0.3 + 3 \times 0.05 = 0.45 \, \mu F/km$$
$$= 20 \times 0.45 = 9.0 \, \mu F \qquad \text{for 20 km}$$

5.12 A single-core cable, consisting of a 1-cm-diameter cable inside a 2.5-cm-diameter sheath, is 10 km long and operates at 13.2 kV and 60 Hz. The relative permittivity of the dielectric is 5, and the open-circuit power factor of the cable is 0.08. Calculate the capacitance of the cable and the charging current through the capacitance.

From (5.11) with $\epsilon = \epsilon_0 \epsilon_r$, we have

$$C = \frac{2\pi\epsilon_0\epsilon_r}{\ln(R_1/R_2)} = \frac{2\pi(10^{-9}/36\pi)5}{\ln(2.5/1)} \, 10 \times 10^3 = 3.03 \, \mu F$$

Charging current $= \omega C V = (377)(3.03 \times 10^{-6})(13.2 \times 10^3) = 15.08 \, A$

5.13 For the cable of Problem 5.12, determine the insulation resistance and the dielectric loss.

From Fig. 5-7(b) and the given data, $\phi = \cos^{-1} 0.08 = 85.4°$. Now, since $\tan \phi = \omega CR$, we have

$$R = \frac{\tan 85.4°}{377 \times 3.03 \times 10^{-6}} = 10.88 \text{ k}\Omega$$

and

$$\text{Dielectric loss} = \frac{V^2}{R} = \frac{13.2^2 \times 10^6}{10.88 \times 10^3} = 16.01 \text{ kW}$$

Supplementary Problems

5.14   A single-core underground cable has a copper conductor of diameter 1.2 cm and resistivity $1.72 \times 10^{-8} \Omega \cdot \text{m}$; a sheath of internal diameter 2.0 cm; and a dielectric (insulating) material of resistivity $5.8 \times 10^{12} \Omega \cdot \text{m}$ and relative permittivity 4. Calculate (a) the conductor resistance and (b) the insulation resistance of a 5-km length of this cable.

   Ans.   (a) 0.76Ω; (b) $94.3 \text{ M}\Omega$

5.15   Determine the capacitance between the core and the sheath of the cable of Problem 5.14 if the permittivity of the dielectric is $27 \times 10^{-12} \text{ F/m}$.

   Ans. $0.332 \mu\text{F}$

5.16   A single-core cable is to operate at 33 kV. If the maximum allowable potential gradient is 4000 kV/m, determine the radius of the conductor and the inner radius of the sheath of an optimally designed cable.

   Ans.   1.17 cm; 3.17 cm

5.17   For the cable shown in Fig. 5-1, $R_1 = 6$ cm, $R_2 = 2$ cm, and relative permittivity $\epsilon_r = 5$. If the cable operates at 33 kV, determine the maximum and minimum values of the electric field strength within the cable.

   Ans.   1504.9 kV/m; 500.6 kV/m

5.18   For the cable shown in Fig. 5-2, $R_1 = 6$ cm, $R_3 = 2$ cm, and $\epsilon_{r2} = 5$ and $\epsilon_{r1} = 4$ are the relative permittivities of the two dielectric layers. Calculate the thickness of each dielectric layer if they are both to sustain the same electric field strength.

   Ans.   2.8 cm; 1.2 cm

5.19   In a certain test on a three-conductor cable, a capacitance C_3 is measured between two conductors, with the third conductor connected to the sheath. Determine C_n of Fig. 5-6(c).

   Ans. $C_n = 2C_3$

5.20   The capacitance between any two conductors of a three-conductor cable, with the third conductor grounded, is 0.6 μF/km. Calculate the line-to-ground capacitance of a 25-km length of the cable.

   Ans.   30 μF

5.21   A single-core cable of conductor radius R_0 has two intersheaths at radii R_1 and R_2 such that the electrical stress varies between the same maximum and minimum in each of the three layers of dielectric. The radius of the sheath is R_3. Obtain a relationship among the four radii.

   Ans. $R_1/R_0 = R_2/R_1 = R_3/R_2$

5.22   The cable of Problem 5.21 operates at 66 kV in a three-phase system and has $R_0 = 1$ cm and $R_3 = 2.565$ cm. Determine the maximum and minimum electrical stresses in the cable.

   Ans.   39.1 kV/cm; 28.3 kV/cm

5.23   Calculate the rms voltage between the two sheaths of the cable of Problem 5.22.

   Ans.   17.4 kV

5.24   The capacitance between any two conductors of a three-phase, three-conductor cable is $2 \, \mu F$. The cable operates at 11 kV line voltage and 50 Hz. What is the charging current through the cable capacitance?

   Ans.   7.98 A

5.25   A three-phase, three-conductor cable, operating at 10 kV line voltage and 25 Hz, has $3 \, \mu F$ capacitance between any two of its conductors. The cable supplies an inductive load taking 30 A of current at a power factor of 0.9 lagging. Determine the sending-end current and power factor.

   Ans.   28.1 A; 0.96 lagging

5.26   What is the conductor diameter of an optimally designed single-core cable operating at 85 kV with a permissible dielectric stress of 6000 kV/m?

   Ans.   2.83 cm

5.27   For a cable of the type shown in Fig. 5-2, $R_3 = 0.5$ cm, $R_1 = 2.5$ cm, $\epsilon_1 = 2.5\epsilon_0$, and $\epsilon_2 = 4\epsilon_0$. The electrical stresses in the inner and outer dielectrics are not to exceed 60 kV/cm and 50 kV/cm, respectively. What is the maximum permissible operating voltage for the cable?

   Ans.   65 kV

Chapter 6

Fault Calculations

The operation of a power system departs from normal after the occurrence of a fault. Faults give rise to abnormal operating conditions—usually excessive currents and voltages at certain points on the system—which are guarded against with various types of protective equipment.

6.1 TYPES OF FAULTS

Various types of short-circuit faults that can occur on a transmission line are depicted in Fig. 6-1; the frequency of occurrence decreases from part (a) to part (f). Although the balanced three-phase short circuit in Fig. 6-1(d) is relatively uncommon, it is the most severe fault and therefore determines the rating of the line-protecting circuit breaker. A *fault study* includes the following:

1. Determination of the maximum and minimum three-phase short-circuit currents

2. Determination of unsymmetrical fault currents, as in single line-to-ground, double line-to-ground, line-to-line, and open-circuit faults

3. Determination of the ratings of required circuit breakers

4. Investigation of schemes of protective relaying

5. Determination of voltage levels at strategic points during a fault

The short-circuit faults depicted in Fig. 6-1 are called *shunt faults*; open circuits, which may be caused by broken conductors, for instance, are categorized as *series faults*.

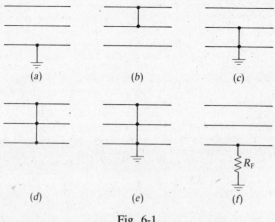

Fig. 6-1.

6.2 SYMMETRICAL FAULTS

A balanced three-phase short circuit [Fig. 6-1(d)] is an example of a symmetrical fault. Balanced three-phase fault calculations can be carried out on a per-phase basis, so that only single-phase equivalent circuits need be used in the analysis. Invariably, the circuit constants are expressed in per-unit terms, and all calculations are made on a per-unit basis. In short-circuit calculations, we often evaluate the short-circuit MVA (megavolt-amperes), which is equal to $\sqrt{3}V_l I_f$, where V_l is the nominal line voltage in kilovolts, and I_f is the fault current in kiloamperes.

An example of a three-phase symmetrical fault is a sudden short at the terminals of a synchronous generator. The symmetrical trace of a short-circuited stator-current wave is shown in Fig. 6-2. The wave, whose envelope is shown in Fig. 6-3, may be divided into three periods or time regimes: the *subtransient period*, lasting only for the first few cycles, during which the current decrement is very rapid; the *transient period*, covering a relatively longer time during which the current decrement is more moderate; and finally the *steady-state period*. The difference $\Delta i'$ (in Fig. 6-3) between the transient envelope and the steady-state amplitude is plotted on a logarithmic scale as a function of time in Fig. 6-4, along with the difference $\Delta i''$ between the subtransient envelope and an extrapolation of the transient envelope. Both plots closely approximate straight lines, illustrating the essentially exponential nature of the decrement.

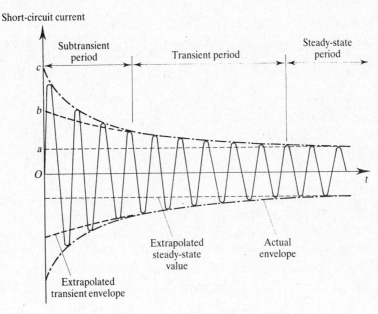

Fig. 6-2.

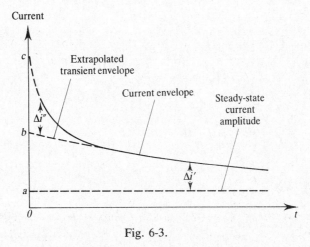

Fig. 6-3.

The currents during these three regimes are limited primarily by various reactances of the synchronous machine (we neglect the armature resistance, which is relatively small). These currents and reactances are defined by the following equations, provided the alternator was operating at no

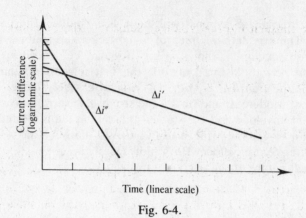

Fig. 6-4.

load before the occurrence of a three-phase fault at its terminals:

$$|I| = \frac{Oa}{\sqrt{2}} = \frac{|E_g|}{X_d} \qquad (6.1)$$

$$|i'| = \frac{Ob}{\sqrt{2}} = \frac{|E_g|}{X'_d} \qquad (6.2)$$

$$|i''| = \frac{Oc}{\sqrt{2}} = \frac{|E_g|}{X''_d} \qquad (6.3)$$

where $|E_g|$ is the no-load voltage of the generator, the currents are rms currents, and O, a, b, and c are shown in Fig. 6-2. The machine reactances X_s, X'_d, and X''_d are known as the *direct-axis synchronous reactance*, *direct-axis transient reactance*, and *direct-axis subtransient reactance*, respectively. The currents I, i', and i'' are known as the steady-state, transient, and subtransient currents. From (6.1) through (6.3) it follows that the fault currents in a synchronous generator can be calculated when the machine reactances are known.

Suppose now that a generator is loaded when a fault occurs. Figure 6-5(a) shows the corresponding equivalent circuit with the fault to occur at point P. The current flowing before the fault occurs is I_L, the voltage at the fault is V_f, and the terminal voltage of the generator is V_t. When a three-phase fault occurs at P, the circuit shown in Fig. 6-5(b) becomes the appropriate equivalent circuit (with switch S closed). Here a voltage E''_g in series with X''_d supplies the steady-state current I_L when switch S is open, and supplies the current to the short circuit through X''_d and Z_{ext} when switch S is closed. If we can determine E''_g, we can find this current through X''_d, which will be i''. With switch S open, we have

$$E''_g = V_t + jI_L X''_d \qquad (6.4)$$

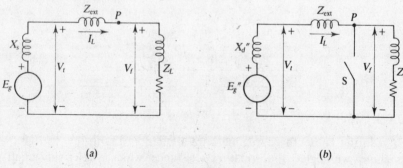

(a)   (b)

Fig. 6-5.

which defines E_g'', the subtransient internal voltage. Similarly, for the transient internal voltage we have

$$E_g' = V_t + jI_L X_d' \qquad (6.5)$$

Clearly E_g'' and E_g' are dependent on the value of the load before the fault occurs.

6.3 UNSYMMETRICAL FAULTS AND SYMMETRICAL COMPONENTS

Unsymmetrical faults such as line-to-line and line-to-ground faults (which occur more frequently than three-phase short circuits) can be analyzed on a per-phase basis. For such faults the method of symmetrical components is used. This method is based on the fact that a set of three-phase unbalanced phasors can be resolved into three sets of symmetrical components, which are termed the *positive-sequence, negative-sequence,* and *zero-sequence components*. The phasors of the set of positive-sequence components have a counterclockwise phase rotation (or phase sequence) *abc*; the negative-sequence components have the reverse phase sequence *acb*; and the zero-sequence components are all in phase with each other. These sequence components are represented geometrically in Fig. 6-6. The positive-sequence components are designated with the subscript 1, and the subscripts 2 and 0 are used for negative- and zero-sequence components, respectively.

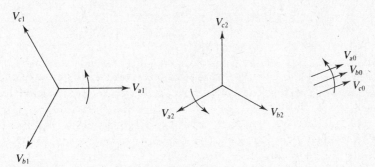

Fig. 6-6.

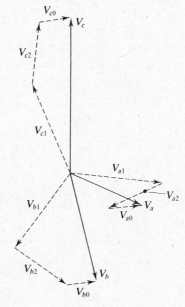

Fig. 6-7.

Thus, the unbalanced system of Fig. 6-7 can be resolved into symmetrical components as shown in Fig. 6-6. In particular, we have

$$V_a = V_{a0} + V_{a1} + V_{a2} \tag{6.6}$$

$$V_b = V_{b0} + V_{b1} + V_{b2} \tag{6.7}$$

$$V_c = V_{c0} + V_{c1} + V_{c2} \tag{6.8}$$

We now introduce an operator a that causes a counterclockwise rotation of 120° (just as the j operator produces a 90° rotation), such that

$$a = 1\underline{/120°} = 1 \times e^{j120} = -0.5 + j0.866$$

$$a^2 = 1\underline{/240°} = -0.5 - j0.866 = a^*$$

$$a^3 = 1\underline{/360°} = 1\underline{/0°}$$

$$1 + a + a^2 = 0$$

Using these properties, we may write the components of a given sequence in terms of any chosen component. From Fig. 6-6, we have

$$V_{b1} = a^2 V_{a1}$$

$$V_{c1} = a V_{a1}$$

$$V_{b2} = a V_{a2}$$

$$V_{c2} = a^2 V_{a2}$$

$$V_{a0} = V_{b0} = V_{c0}$$

Consequently, (6.6) to (6.8) become, in terms of components of phase a,

$$V_a = V_{a0} + V_{a1} + V_{a2} \tag{6.9}$$

$$V_b = V_{a0} + a^2 V_{a1} + a V_{a2} \tag{6.10}$$

$$V_c = V_{a0} + a V_{a1} + a^2 V_{a2} \tag{6.11}$$

Solving for the sequence components from (6.9) through (6.11) yields

$$V_{a0} = \tfrac{1}{3}(V_a + V_b + V_c) \tag{6.12}$$

$$V_{a1} = \tfrac{1}{3}(V_a + a V_b + a^2 V_c) \tag{6.13}$$

$$V_{a2} = \tfrac{1}{3}(V_a + a^2 V_b + a V_c) \tag{6.14}$$

Equations similar to (6.9) to (6.14) hold for currents as well.

A quantity (current, voltage, impedance, power) that is given in terms of symmetrical components is sometimes called the *sequence* quantity, as in "sequence current."

6.4 SEQUENCE POWER

To obtain the power in a three-phase system in terms of symmetrical components, we rewrite (6.9) through (6.14) in matrix notation as follows:

$$\mathbf{V} = \mathbf{A}\mathbf{V}' \tag{6.15}$$

where
$$\mathbf{V} = \begin{bmatrix} V_a \\ V_b \\ V_c \end{bmatrix} \qquad \mathbf{V}' = \begin{bmatrix} V_{a0} \\ V_{a1} \\ V_{a2} \end{bmatrix} \qquad \mathbf{A} = \begin{bmatrix} 1 & 1 & 1 \\ 1 & a^2 & a \\ 1 & a & a^2 \end{bmatrix}$$

Similarly, for the currents we have

$$\mathbf{I} = \mathbf{A}\mathbf{I}' \tag{6.16}$$

where

$$\mathbf{I} = \begin{bmatrix} I_a \\ I_b \\ I_c \end{bmatrix} \quad \text{and} \quad \mathbf{I}' = \begin{bmatrix} I_{a0} \\ I_{a1} \\ I_{a2} \end{bmatrix}$$

The average complex power S may now be written as

$$S = \tilde{\mathbf{V}}\mathbf{I}^* \tag{6.17}$$

where $\tilde{\mathbf{V}}$ is the transpose of $\mathbf{V}$, and $\mathbf{I}^*$ is the complex conjugate of $\mathbf{I}$. From (6.15) and (6.16), we have

$$\tilde{\mathbf{V}}' = \tilde{\mathbf{V}}'\tilde{\mathbf{A}} \tag{6.18}$$

and

$$\mathbf{I}' = \mathbf{A}^*\mathbf{I}'^* \tag{6.19}$$

Consequently, (6.17) through (6.19) yield

$$S = \tilde{\mathbf{V}}'\tilde{\mathbf{A}}\mathbf{A}^*\mathbf{I}'^* \tag{6.20}$$

Now, since

$$\mathbf{A}\mathbf{A}^* = \begin{bmatrix} 1 & 1 & 1 \\ 1 & a^2 & a \\ 1 & a & a^2 \end{bmatrix}\begin{bmatrix} 1 & 1 & 1 \\ 1 & a & a^2 \\ 1 & a^2 & a \end{bmatrix} = \begin{bmatrix} 3 & 0 & 0 \\ 0 & 3 & 0 \\ 0 & 0 & 3 \end{bmatrix} \tag{6.21}$$

(6.20) becomes

$$S = V_a I_a^* + V_b I_b^* + V_c I_c^* = 3(V_{a0}I_{a0}^* + V_{a1}I_{a1}^* + V_{a2}I_{a2}^*) \tag{6.22}$$

Thus, the sequence power is one-third the power in terms of phase quantities.

6.5 SEQUENCE IMPEDANCES AND SEQUENCE NETWORKS

Corresponding to sequence currents, we may define sequence impedances. An impedance through which only positive-sequence currents flow is called the *positive-sequence impedance*. Similarly, when only negative-sequence currents flow, the impedance is known as the *negative-sequence impedance*; and when zero-sequence currents alone are present, the impedance is called the *zero-sequence impedance*.

Unsymmetrical (or unbalanced) fault calculations are facilitated by the use of the concepts of sequence voltages, currents, and impedances. Because a voltage of a specific sequence produces a current of the same sequence only, the various sequence networks representing an unbalanced condition have no mutual coupling. This feature of sequence networks simplifies the calculations considerably.

Solved Problems

6.1   Figure 6-8 shows the one-line diagram for a single phase of a system in which a generator supplies a load through a step-up transformer, a transmission line, and a step-down transformer. Calculate the per-unit current. The transformers are ideal.

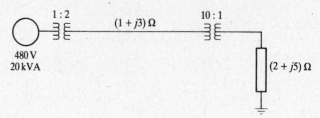

Fig. 6-8.

Because the voltage (and current) levels change across the transformers, different base voltages are involved at different parts of the system. At the generator,

$$V_{base,gen} = 480\,V = 1\,pu$$

$$kVA_{base,gen} = 20\,kVA = 1\,pu$$

$$I_{base,gen} = \frac{20{,}000}{480} = 41.67\,A = 1\,pu$$

$$Z_{base,gen} = \frac{480}{41.67} = 11.52\,\Omega = 1\,pu$$

Along the transmission line,

$$V_{base,line} = \frac{480}{0.5} = 960\,V = 1\,pu$$

$$kVA_{base,line} = 20\,kVA = 1\,pu$$

$$I_{base,line} = \frac{20{,}000}{960} = 20.83 = 1\,pu$$

$$Z_{base,line} = \frac{960}{20.83} = 46.08\,\Omega = 1\,pu$$

$$Z_{line} = \frac{1 + j3}{46.08} = (0.022 + j0.065)\,pu$$

At the load,

$$V_{base,load} = \frac{960}{10} = 96\,V = 1\,pu$$

$$kVA_{base,load} = 20\,kVA = 1\,pu$$

$$I_{base,load} = \frac{20{,}000}{96} = 208.3 = 1\,pu$$

$$Z_{base,load} = \frac{96}{208.3} = 0.4608\,\Omega = 1\,pu$$

$$Z_{load} = \frac{2 + j5}{0.4608} = (4.34 + j10.85)\,pu$$

The total impedance is then

$$Z_{total} = Z_{line} + Z_{load} = (0.022 + j0.065) + (4.34 + j10.85)$$
$$= 4.362 + j10.915 = 11.75\underline{/68°}\,pu$$

so that
$$I = \frac{1\underline{/0°}}{Z_{total}} = \frac{1\underline{/0°}}{11.75\underline{/68°}} = 0.085\underline{/-68°}\,pu$$

6.2   An interconnected generator-reactor system is shown in Fig. 6-9(a). The base values for the

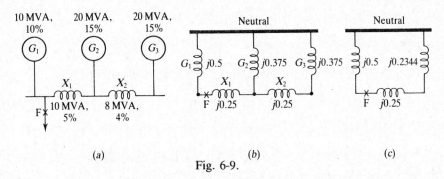

(a)　　　　　　　　　　　　　　　(b)　　　　　　　　　　　　　(c)

Fig. 6-9.

given percent reactances are the ratings of the individual pieces of equipment. A three-phase short-circuit occurs at point F. Determine the fault current and the fault kVA if the busbar line-to-line voltage is 11 kV.

First we arbitrarily choose 50 MVA as the base MVA, and find per-unit values for the system reactances, referred to this base. We obtain

$$X_{G1} = \frac{50}{10}0.10 = 0.5 \text{ pu}$$

$$X_{G2} = \frac{50}{20}0.15 = 0.375 \text{ pu}$$

$$X_{G3} = \frac{50}{20}0.15 = 0.375 \text{ pu}$$

$$X_1 = \frac{50}{10}0.5 = 0.25 \text{ pu}$$

$$X_2 = \frac{50}{8}0.4 = 0.25 \text{ pu}$$

These reactances produce the per-phase reactance diagram of Fig. 6-9(b), which is simplified to Fig. 6-9(c). The total reactance from the neutral to the fault at F is, from that diagram,

$$\text{Per-unit reactance} = j\frac{0.5(0.2344 + 0.25)}{0.5 + (0.2344 + 0.25)} = j0.246$$

Then

$$\text{Fault MVA} = \frac{50 \times 10^3}{0.246} = 203.25 \text{ MVA}$$

and

$$\text{Fault current} = \frac{203.25 \times 10^6}{\sqrt{3} \times 11 \times 10^3} = 10,668 \text{ A}$$

6.3　A three-phase short-circuit fault occurs at point F in the system shown in Fig. 6-10(a). Calculate the fault current.

Let the base MVA be 30 MVA, and let 33 kV be the base voltage. Then referred to these values, we have the following reactances and impedance:

$$X_{G1} = \frac{30}{20}0.15 = 0.225 \text{ pu}$$

$$X_{G2} = \frac{30}{10}0.10 = 0.3 \text{ pu}$$

$$X_{\text{trans}} = \frac{30}{30}0.005 = 0.5 \text{ pu}$$

$$Z_{\text{line}} = (3 + j15)\frac{30}{33^2}$$

$$= (0.0826 + j0.4132) \text{ pu}$$

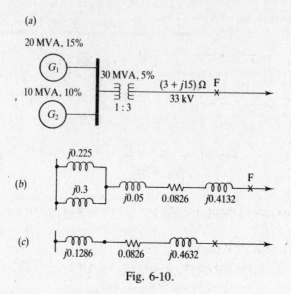

Fig. 6-10.

These per-unit values are shown in Fig. 6-10(b), which can be reduced to Fig. 6-10(c). From that diagram we find that the total impedance from the generator neutral to the fault is

$$\text{Per-unit } Z_{total} = 0.0826 + j0.5918 = 0.5975\underline{/82°} \text{ pu}$$

Then

$$\text{Short-circuit kVA} = \frac{30 \times 10^3}{0.5975} = 50.21 \text{ MVA}$$

and

$$\text{Short-circuit current} = \frac{50.21 \times 10^6}{\sqrt{3} \times 33 \times 10^3} = 878.5 \text{ A}$$

6.4   The phase currents in a wye-connected, unbalanced load are $I_a = (44 - j33)$, $I_b = -(32 + j24)$, and $I_c = (-40 + j25)$ A. Determine the sequence currents.

From (6.12) through (6.14), adapted for currents, we obtain

$$I_{a0} = \tfrac{1}{3}[(44 - j33) - (32 + j24) + (-40 + j25)]$$
$$= -9.33 - j10.67 = 14.17\underline{/-131.2°} \text{ A}$$
$$I_{a1} = \tfrac{1}{3}[(44 - j33) - (-0.5 + j0.866)(32 + j24) - (0.5 - j0.866)(-40 + j25)]$$
$$= 40.81 - j8.77 = 41.74\underline{/-12.1°} \text{ A}$$
$$I_{a2} = \tfrac{1}{3}[(44 - j33) + (0.5 - j0.866)(32 + j24) + (-0.5 + j0.866)(-40 + j25)]$$
$$= 12.52 - j13.48 = 18.37\underline{/-47°} \text{ A}$$

6.5   A three-phase, wye-connected load is connected across a three-phase, balanced supply system. Obtain a set of equations relating the symmetrical components of the line and phase voltages.

The symmetrical system, the assumed directions of voltages, and the nomenclature are shown in Fig. 6-11, from which we have

$$V_{ab} = V_a - V_b \qquad V_{bc}' = V_b - V_c \qquad V_{ca} = V_c - V_a$$

Because $V_{ab} + V_{bc} + V_{ca} = 0$, we get

$$V_{ab0} = V_{bc0} = V_{ca0} = 0$$

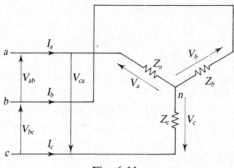

Fig. 6-11.

We choose V_{ab} as the reference phasor. For the positive-sequence component, we have

$$
\begin{aligned}
V_{ab1} &= \tfrac{1}{3}(V_{ab} + aV_{bc} + a^2 V_{ca}) \\
&= \tfrac{1}{3}[(V_a - V_b) + a(V_b - V_c) + a^2(V_c - V_a)] \\
&= \tfrac{1}{3}[(V_a + aV_b + a^2 V_c) - (a^2 V_a + V_b + aV_c)] \\
&= \tfrac{1}{3}[(V_a + aV_b + a^2 V_c) - a^2(V_a + aV_b + a^2 V_c)] \\
&= \tfrac{1}{3}[(1 - a^2)(V_a + aV_b + a^2 V_c)] = (1 - a^2)V_{a1} \\
&= \sqrt{3}V_{a1}e^{j30^\circ}
\end{aligned}
\tag{1}
$$

Similarly, for the negative-sequence component, we obtain

$$
\begin{aligned}
V_{ab2} &= \tfrac{1}{3}(V_{ab} + a^2 V_{bc} + aV_{ca}) \\
&= \tfrac{1}{3}[(V_a - V_b) + a^2(V_b - V_c) + a(V_c - V_a)] \\
&= \tfrac{1}{3}[(V_a + a^2 V_b + aV_c) - (aV_a + V_b + a^2 V_c)] \\
&= \tfrac{1}{3}[(V_a + a^2 V_b + aV_c) - a(V_a + a^2 V_b + aV_c)] \\
&= \tfrac{1}{3}(1 - a)(V_a + a^2 V_b + aV_c) = (1 - a)V_{a2} \\
&= \sqrt{3}V_{a2}e^{-j30^\circ}
\end{aligned}
\tag{2}
$$

In (1) and (2), V_{a1} and V_{a2} are, respectively, the positive- and negative-sequence components of the phase voltage V_a.

Proceeding as in the derivation of (1) and (2), but now choosing V_{bc} as the reference phasor, we can show that

$$
V_{bc1} = -j\sqrt{3}V_{a1}
\tag{3}
$$

$$
V_{bc2} = j\sqrt{3}V_{a2}
\tag{4}
$$

6.6   The line voltages across a three-phase, wye-connected load, consisting of a 10-Ω resistance in each phase, are unbalanced such that $V_{ab} = 220\underline{/131.7^\circ}$ V, $V_{bc} = 252\underline{/0^\circ}$ V, and $V_{ca} = 195\underline{/-122.6^\circ}$ V. Determine the sequence phase voltages. Then find the voltages across the 10-Ω resistances, and calculate the line currents.

Since the line voltages are given, we first determine the sequence components of the line voltages. Using (6.13) and (6.14) for line voltages yields

$$
\begin{aligned}
V_{bc1} &= \tfrac{1}{3}(V_{bc} + aV_{ca} + a^2 V_{ab}) = \tfrac{1}{3}(252\underline{/0^\circ} + 1\underline{/120^\circ} \times 195\underline{/-122.6^\circ} + 1\underline{/-120^\circ} \times 220\underline{/131.7^\circ}) \\
&= 221 + j12 \text{ V} \\
V_{bc2} &= \tfrac{1}{3}(V_{bc} + a^2 V_{ca} + aV_{ab}) = \tfrac{1}{3}(252\underline{/0^\circ} + 1\underline{/-120^\circ} \times 195\underline{/-122.6^\circ} + 1\underline{/120^\circ} \times 220\underline{/131.7^\circ}) \\
&= 31 - j11.9
\end{aligned}
$$

From (6.12) we have

$$V_{bc0} = \tfrac{1}{3}(V_{bc} + V_{ca} + V_{ab}) = \tfrac{1}{3}(252\underline{/0°} + 195\underline{/-122.6°} + 220\underline{/131.7°})$$
$$= 0 \text{ V}$$

The sequence components of the phase voltages are then $V_{a0} = 0$ and, from (3) and (4) of Problem 6.5,

$$V_{a1} = \frac{V_{bc1}}{-j\sqrt{3}} = \frac{221 + j12}{-j\sqrt{3}} = (-6.9 + j127.5) \text{ V}$$

$$V_{a2} = \frac{V_{bc2}}{j\sqrt{3}} = \frac{31 - j11.9}{j\sqrt{3}} = (-6.9 - j17.9) \text{ V}$$

Hence, from (6.9) and (6.10), after simplification,

$$V_a = -6.9 + j127.5 - 6.9 - j17.9 = (-13.8 + j10.96) \text{ V}$$
$$V_b = a^2 V_{a1} + a V_{a2} = (132.8 - j54.8) \text{ V}$$

The line currents I_a and I_b are given by

$$I_a = \frac{V_a}{R} = \tfrac{1}{10}(-13.8 + j109.6) = -1.38 + j10.96 = 11.05\underline{/97.2°} \text{ A}$$

and

$$I_b = \frac{V_b}{R} = \tfrac{1}{10}(132.8 - j54.8) = 13.28 - j5.48 = 14.37\underline{/-22.4°} \text{ A}$$

Since $I_a + I_b + I_c = 0$, we also obtain

$$I_c = -I_a - I_b = -1.38 + j10.96 + 13.28 - j5.48$$
$$= -11.9 - j5.48 = 13.1\underline{/-155.3°} \text{ A}$$

6.7　　A three-phase synchronous generator, grounded through an impedance Z_n, is shown in Fig.

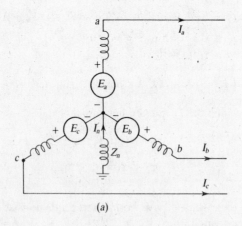

(a)

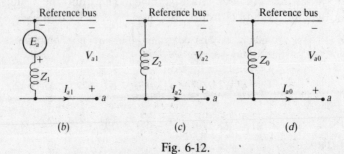

(b)　　　　　　　　　(c)　　　　　　　　　(d)

Fig. 6-12.

6.12(a). The generator is not supplying load, but because of a fault at the generator terminals, currents I_a, I_b, and I_c flow through phases a, b, and c, respectively. Develop and draw the *sequence networks* for the generator for this condition.

Let the generator-induced voltages in the three phases be E_a, E_b, and E_c. The induced voltages in the generator are balanced. Therefore, these voltages are of positive sequence only. For the positive-sequence (phase) voltage, we have

$$V_{a1} = E_a - I_{a1}Z_1 \tag{1}$$

where $I_{a1}Z_1$ is the positive-sequence voltage drop in the positive-sequence impedance Z_1 of the generator. If Z_2 is the negative-sequence impedance of the generator, the negative-sequence voltage at the terminal of phase a is simply

$$V_{a2} = -I_{a2}Z_2 \tag{2}$$

since there is no negative-sequence generated voltage. The generator zero-sequence currents flow through Z_n as well as through Z_{g0}, the generator zero-sequence impedance. The total zero-sequence current through Z_n is $I_{a0} + I_{b0} + I_{c0} = 3I_{a0}$, but the current through Z_{g0} is I_{a0}. Hence

$$V_{a0} = -I_{a0}Z_{g0} - 3I_{a0}Z_n = -I_{a0}(Z_{g0} + 3Z_n) = -I_{a0}Z_0 \tag{3}$$

where $Z_0 = Z_{g0} + 3Z_n$. Equations (1), (2), and (3) are respectively represented by Fig. 6-12(b), (c), and (d).

6.8 A line-to-ground fault occurs on phase a of the generator of Fig. 6-12(a), which was operating without a load. Derive a sequence-network representation of this condition, and determine the current in phase a.

The constraints corresponding to the fault are $I_b = I_c = 0$ (lines remain open-circuited) and $V_a = 0$ (line-to-ground short-circuit). Consequently, the symmetrical components of the current in phase a are

$$I_{a0} = \tfrac{1}{3}(I_a + I_b + I_c) = \tfrac{1}{3}I_a$$
$$I_{a1} = \tfrac{1}{3}(I_a + aI_b + a^2I_c) = \tfrac{1}{3}I_a$$
$$I_{a2} = \tfrac{1}{3}(I_a + a^2I_b + aI_c) = \tfrac{1}{3}I_a$$

so that
$$I_{a0} = I_{a1} = I_{a2} = \tfrac{1}{3}I_a$$

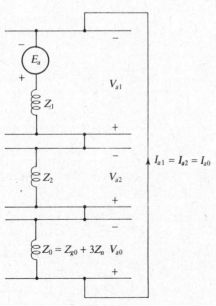

Fig. 6-13.

Hence the sequence networks must be connected in series, as shown in Fig. 6-13. The sequence voltages appear as marked in the figure.

To determine the current I_a, we write, from Fig. 6-13,

$$V_{a0} + V_{a1} + V_{a2} = E_a - I_{a1}Z_1 - I_{a1}Z_2 - I_{a1}Z_0$$

But since

$$V_a = V_{a0} + V_{a1} + V_{a2} = 0$$

we have

$$I_{a1} = \frac{E_a}{Z_1 + Z_2 + Z_0} = \tfrac{1}{3}I_a \qquad (1)$$

and

$$I_a = \frac{3E_a}{Z_1 + Z_2 + Z_0}$$

6.9 A short circuit occurs between phases b and c of a solidly grounded unloaded generator, as shown in Fig. 6-14(a). Obtain a sequence network for this operating condition.

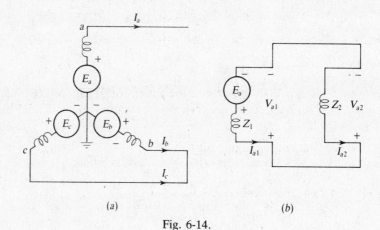

(a)   (b)

Fig. 6-14.

From Fig. 6-14(a), the current and voltage constraints are

$$I_a = 0 \qquad \text{(line open)} \qquad (1)$$

$$I_b + I_c = 0 \qquad \text{(line-to-line short circuit)} \qquad (2)$$

$$V_b = V_c \qquad (3)$$

Substituting (1) and (2) in (6.12) through (6.14) expressed as currents, we obtain

$$I_{a0} = \tfrac{1}{3}(0 + 0) = 0 \qquad (4)$$

$$I_{a1} = \tfrac{1}{3}(0 + aI_b - a^2I_b) = \tfrac{1}{3}(a - a^2)I_b \qquad (5)$$

$$I_{a2} = \tfrac{1}{3}(0 + a^2I_b - aI_b) = \tfrac{1}{3}(a^2 - a)I_b \qquad (6)$$

From (5) and (6) we observe that

$$I_{a1} = -I_{a2} \qquad (7)$$

whereas (4) shows that the zero-sequence current is absent.

Now, from (3) (6.10), and (6.11), we obtain

$$V_{a0} + a^2V_{a1} + aV_{a2} = V_{a0} + aV_{a1} + a^2V_{a2}$$

Hence,

$$V_{a1} = V_{a2} \qquad (8)$$

In terms of sequence impedances (8) may be written as

$$E_a - I_{a1}Z_1 = -I_{a2}Z_2 \qquad (9)$$

Combining (7) and (9) and solving for I_{a1} yield

$$I_{a1} = \frac{E_a}{Z_1 + Z_2} \tag{10}$$

Equations (7) through (10) may be represented by the sequence network in Fig. 6-14(b).

6.10   Develop the sequence network for an unloaded generator with a double line-to-ground fault as shown in Fig. 6-15(a).

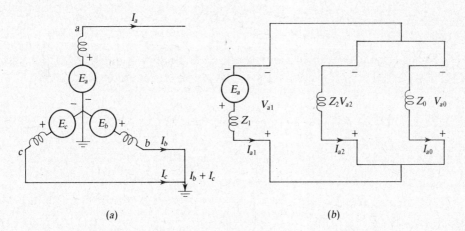

(a)   (b)

Fig. 6-15.

For this case, the current and voltage constraints are

$$I_a = 0 \tag{1}$$

$$V_b = V_c = 0 \tag{2}$$

Proceeding as in Problem 6.9, we use (6.12) to (6.14) to find, for the sequence components of the voltages,

$$V_{a0} = V_{a1} = V_{a2} = \tfrac{1}{3}V_a \tag{3}$$

Consequently, the sequence network equations become

$$E_a - I_{a1}Z_1 = -I_{a0}Z_0 = -I_{a2}Z_2 \tag{4}$$

Solving for I_{a0} and I_{a2} from (4), we obtain

$$I_{a0} = -\frac{E_a - I_{a1}Z_1}{Z_0} \quad \text{and} \quad I_{a2} = \frac{E_a - I_{a1}Z_1}{Z_2} \tag{5}$$

From (6.9), (1), and (5) we then obtain

$$I_a = -\frac{E_a - I_{a1}Z_1}{Z_0} + I_{a1} - \frac{E_a - I_{a1}Z_1}{Z_2} = 0 \tag{6}$$

from which

$$I_{a1} = \frac{E_a}{Z_1 + \dfrac{Z_0 Z_2}{Z_0 + Z_2}} \tag{7}$$

The denominator of (7) shows that Z_0 and Z_2 are connected in parallel, and this parallel combination is connected in series with Z_1. Hence, the sequence network representing (7) is that shown in Fig. 6-15(b).

6.11   The positive-, negative-, and zero-sequence reactances of a 20-MVA, 13.2-kV synchronous generator are 0.3 pu, 0.2 pu, and 0.1 pu, respectively. The generator is solidly grounded and is not loaded. A line-to-ground fault occurs on phase a. Neglecting all the resistances, determine the fault current.

The sequence network corresponding to this fault is shown in Fig. 6-13. Let $E_a = 1\underline{/0^\circ}$ pu. Since the total reactance is $j0.3 + j0.2 + j0.1 = j0.6$, from ($1$) of Problem 6.8 we have

$$I_{a1} = \frac{1\underline{/0^\circ}}{j0.6} = 1.67\underline{/-90^\circ} = -j1.67 \text{ pu}$$

and
$$I_a = 3I_{a1} = 5\underline{/-90^\circ} \text{ pu}$$

Choosing the rated values as base quantities, we have

$$\text{Base current} = \frac{20,000}{\sqrt{3} \times 13.2} = 874.8 \text{ A} = 1 \text{ pu}$$

$$\text{Fault current} = I_a = 5 \text{ pu} = 5 \times 874.8 = 4374 \text{ A}$$

6.12   A line-to-line fault occurs at the terminals of the unloaded generator of Problem 6.11. Calculate the fault current.

For this fault condition the sequence network is that shown in Fig. 6-14(b). Let $E_a = 1\underline{/0^\circ}$ pu. Then, from (4), (7), and (10) of Problem 6.9 we obtain

$$I_{a0} = 0$$

$$I_{a1} = -I_{a2} = \frac{1\underline{/0^\circ}}{j0.3 + j0.2} = 2\underline{/-90^\circ} = -j2.0 \text{ pu}$$

Hence, the fault current is given by

$$I_b = I_{b0} + I_{b1} + I_{b2} = 0 + a^2 I_{a1} + a I_{a2}$$
$$= (-0.5 - j0.866)(-j2.0) + (-0.5 + j0.866)(j2.0)$$
$$= (j1.0 - 1.732 - j1.0 - 1.732) = -3.464\underline{/0^\circ} \text{ pu}$$

As calculated in Problem 6.11, the base current is 874.8 A. Hence,

$$\text{Fault current} = I_b = 874.8 \times 3.464 = 3030 \text{ A}$$

6.13   The generator of Problem 6.11 is initially unloaded. A double line-to-ground fault occurs at the generator terminals. Calculate the fault current and the line voltages.

The sequence network for this case is shown in Fig. 6-15. We let $E_a = 1\underline{/0^\circ}$ pu. Then, from (7) of Problem 6.10 we get

$$I_{a1} = \frac{1\underline{/0^\circ}}{j0.3 + \dfrac{(j0.2)(j0.1)}{j0.2 + j0.1}} = -j2.73 \text{ pu}$$

Also, from Fig. 6-15,

$$V_{a1} = E_a - I_{a1}Z_1 = 1\underline{/0^\circ} - (-j2.73)(j0.3) = 0.181 \text{ pu}$$

and, from (3) of Problem 6.10, $V_{a2} = V_{a0} = 0.181$ pu. Therefore,

$$I_{a2} = -\frac{V_{a2}}{Z_2} = \frac{-0.181}{j0.2} = j0.905 \text{ pu}$$

$$I_{a0} = -\frac{V_{a0}}{Z_0} = \frac{-0.181}{j0.1} = j1.81 \text{ pu}$$

From Fig. 6-15(a), the fault current is $I_b + I_c$. Expressing this sum in terms of sequence components of I_a, we have

$$I_b + I_c = (I_{a0} + a^2 I_{a1} + a I_{a2}) + (I_{a0} + a I_{a1} + a^2 I_{a2})$$
$$= 2I_{a0} + (a + a^2)(I_{a1} + I_{a2})$$
$$= 2I_{a0} - (I_{a1} + I_{a2}) \tag{1}$$

Since $I_a = 0$ from (1) of Problem 6.10, we may write

$$I_a = I_{a0} + I_{a1} + I_{a2} = 0$$

or

$$-(I_{a1} + I_{a2}) = I_{a0} \tag{2}$$

Now (1) and (2) yield

$$\text{Fault current} = I_b + I_c = 3I_{a0} = 3(j1.81) = 5.43 \, \text{pu}$$
$$= 5.43 \times 874.8 = 4750 \, \text{A}$$

To calculate the line voltages, we use (2) and (3) of Problem 6.10. They yield

$$V_a = 3V_{a1} = 3 \times 0.181 = 0.543 \, \text{pu}$$

and

$$V_b = V_c = 0$$

Hence,

$$V_{ab} = V_a = 0.543 \times \frac{13.2}{\sqrt{3}} = 4.14 \, \text{kV}$$

$$V_{bc} = 0$$

$$V_{ca} = V_a = 0.543 \times \frac{13.2}{\sqrt{3}} = 4.14 \, \text{kV}$$

6.14 Calculate the line-to-line voltages for the generator of Problem 6.12 (which has a line-to-line fault).

To determine the line voltages, we must first determine their sequence components. From Fig. 6-14(b) and Problem 6.12,

$$V_{a1} = E_a - I_{a1}Z_1 = 1\underline{/0°} - (-j2)(j0.3) = 0.4\underline{/0°} \, \text{pu}$$
$$V_{a2} = -I_{a2}Z_2 = (-j2)(j0.2) = 0.4\underline{/0°} = V_{a1}$$
$$V_a = V_{a0} + V_{a1} + V_{a2} = 2 \times 0.4 = 0.8\underline{/0°} \, \text{pu}$$
$$V_b = a^2 V_{a1} + a V_{a2} = (-0.5 - j0.866)(0.4\underline{/0°}) + (-0.5 + j0.866)(0.4\underline{/0°}) = 0.4\underline{/0°} \, \text{pu}$$
$$V_c = V_b = 0.4\underline{/0°}$$

The line voltages are then

$$V_{ab} = V_a - V_b = 0.8 - (-0.4) = 1.2 \, \text{pu} = 1.2 \times \frac{13.2}{\sqrt{3}} = 9.145 \, \text{kV}$$

$$V_{ac} = V_a - V_c = 1.2 \, \text{pu} = 9.145 \, \text{kV}$$
$$V_{bc} = V_b - V_c = 0 \, \text{pu} = 0 \, \text{V}$$

6.15 Determine the voltages V_a, V_b, and V_c for the generator of Problem 6.11.

From Fig. 6-13, we have the per-unit sequence voltages

$$V_{a1} = 1 - (-j1.67)(j0.3) = 0.5 \, \text{pu}$$
$$V_{a2} = -(-j1.67)(j0.2) = -0.333\underline{/0°} \, \text{pu}$$
$$V_{a0} = -(-j1.67)(j0.1) = -0.167\underline{/0°} \, \text{pu}$$

Then

$$V_a = 0.5 - 0.333 - 0.167 = 0 \, \text{pu}$$

Since $V_a = 0$, the line-to-line voltages are

$$V_{ab} = V_a - V_b = -V_b = -0.25 - j0.721 = 0.673\underline{/-109.12°}\ \text{pu}$$

$$V_{bc} = V_b - V_c = -j1.442 = 1.442\underline{/-90°}\ \text{pu}$$

$$V_{ca} = V_c - V_a = V_c = -0.25 + j0.721 = 0.673\underline{/109.12°}\ \text{pu}$$

Since the base voltage is $13.2/\sqrt{3}$, we finally have

$$V_{ab} = V_{ca} = 0.673\left(\frac{13.2}{\sqrt{3}}\right) = 5.13\ \text{kV}$$

and

$$V_{bc} = 1.442\left(\frac{13}{2\sqrt{3}}\right) = 10.99\ \text{kV}$$

To determine V_b and V_c, we determine their sequence components. Thus, we have

$$V_{b1} = a^2 V_{a1} = (-0.5 - j0.866)0.5 = -0.25 - j0.433\ \text{pu}$$

$$V_{b2} = aV_{a2} = (-0.5 + j0.866)(-0.333) = 0.167 - j(0.288)\ \text{pu}$$

$$V_{b0} = V_{a0} = -0.167\ \text{pu}$$

Hence, $$V_b = -0.167 - 0.25 - j0.433 + 0.167 - j0.288 = -0.25 - j0.721\ \text{pu}$$

In similar fashion, we find

$$V_{c0} = V_{a0} = -0.167\ \text{pu}$$

$$V_{c1} = aV_{a1} = (-0.5 + j0.866)0.5 = -0.25 + j0.433\ \text{pu}$$

$$V_{c2} = a_2 V_{a2} = (-0.5 - j0.866)(-0.333) = 0.167 + j0.288\ \text{pu}$$

and $$V_c = -0.167 - 0.25 + j0.433 + 0.167 + j0.288 = -0.25 + j0.721\ \text{pu}$$

6.16 The system shown in Fig. 6-16(*a*) is initially on no load. Calculate the subtransient fault current that results when a three-phase fault occurs at F, given that the transformer voltage on the high-voltage side is 66 kV.

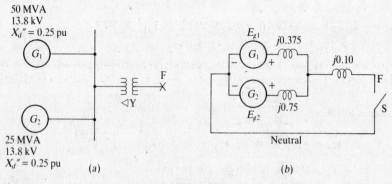

Fig. 6-16.

Let the base voltage (on the high side) be 69 kV, and the base kVA be 75 MVA. Then, in per-unit terms we have, for generator G_1,

$$X''_d = 0.25\left(\frac{75,000}{50,000}\right) = 0.375\ \text{pu}$$

and $$E_{g1} = \frac{66}{69} = 0.957\ \text{pu}$$

For generator G_2,

$$X_d'' = 0.25\left(\frac{75,000}{25,000}\right) = 0.750 \text{ pu}$$

and

$$E_{g2} = \frac{66}{69} = 0.957 \text{ pu}$$

For the transformer, $X = 0.10$ pu.

Figure 6-16(b) shows the reactance diagram for the system before the fault occurs; the fault is simulated by closing switch S. The two parallel subtransient reactances are equivalent to a reactance

$$X_{sub} = \frac{0.375 \times 0.75}{0.375 + 0.75} = 0.25 \text{ pu}$$

Therefore, as a phasor with E_{g1} as reference, the subtransient current in the short circuit is

$$i'' = \frac{0.957}{j0.25 + j0.10} = -j2.735 \text{ pu}$$

6.17 The per-unit reactances of a synchronous generator are $X_d = 1.0$, $X_d' = 0.35$, and $X_d'' = 0.25$. The generator supplies a 1.0 per-unit load at 0.8 power factor lagging. Calculate the voltages behind the synchronous, transient, and subtransient reactances.

With $V_t = 1 + j0$ as base and using

$$E_g = V_t + jI_L X_d$$

as well as (*6.4*) and (*6.5*), we obtain

$$E_g = (1 + j0) + j1.0(0.8 - j0.6) = 1.79 \text{ pu}$$
$$E_g' = (1 + j0) + j0.35(0.8 - j0.6) = 1.24 \text{ pu}$$
$$E_g'' = (1 + j0) + j0.25(0.8 - j0.6) = 1.17 \text{ pu}$$

Supplementary Problems

6.18 A portion of a power system is shown in Fig. 6-17, which also shows the ratings of the generators and the transformer and their respective percent reactances. A symmetrical short circuit appears on a feeder at F. Find the value of the reactance X (in percent) such that the short-circuit MVA does not exceed 300 MVA.

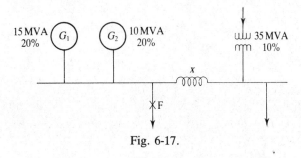

Fig. 6-17.

Ans.   30 percent

6.19 Three generators, each rated at 10 MVA and having a reactance of 10 percent, are connected to a common busbar and supply a load through two 15-kVA step-up transformers. Each transformer has a reactance of 7 percent. Determine the maximum fault MVA on (*a*) the high-voltage side and (*b*) the low-voltage side.

Ans. (*a*) 68.18 MVA; (*b*) 100 MVA

6.20 For the system shown in Fig. 6-18, calculate the short-circuit MVA at A and at B.

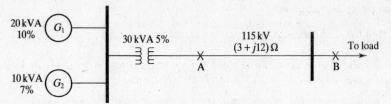

Fig. 6-18.

Ans. 0.218 MVA; 0.218 MVA (approximately)

6.21 The reactance diagram of a portion of a power system is shown in Fig. 6-19. The line-to-ground source voltage is 1.0 pu, and a line-to-ground fault occurs at P. Determine the per-unit currents in the two portions of transmission line B.

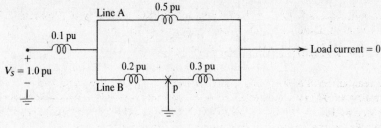

Fig. 6-19.

Ans. 3.077 pu; 0.769 pu

6.22 A three-phase short-circuit occurs at F in the system of Fig. 6-20. Calculate the fault MVA. The reactances are all in percent.

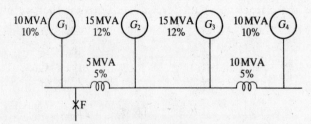

Fig. 6-20.

Ans. 176 MVA

6.23 The line currents in a three-phase, four-wire system are $I_a = (300 + j400)$ A, $I_b = (200 + j200)$ A, and $I_c = (-400 - j200)$ A. Determine the positive-, negative-, and zero-sequence components.

Ans. $276.0/25.6°$ A; $307/86.7°$ A; $137.4/76°$ A

6.24 The line currents in a delta-connected load are $I_a = 5/0°$, $I_b = 7/200°$, and $I_c = 5/90°$ A. Calculate the positive-, negative-, and zero-sequence components of the current for phase a. Also determine the positive- and negative-sequence components of current I_{ab}, and hence calculate I_{ab}.

    Ans. $5.57/-15°$ A; $7.94/224°$ A; 0; $3.21/15°$ A; $4.59/-166°$ A; 1.38 A

6.25   An unbalanced, wye-connected load consisting of the phase resistances $R_a = 60\,\Omega$, $R_b = 40\,\Omega$, and $R_c = 80\,\Omega$ is connected to a 440-V, three-phase, balanced supply. Calculate the line currents by the method of symmetrical components.

    Ans. $4.473/100.67°$ A; $5.138/-34.71°$ A; $3.696/156.63°$ A

6.26   A three-phase, unbalanced delta load draws 100 A of line current from a balanced three-phase supply. An open-circuit fault occurs on one of the lines. Determine the sequence components of the currents in the unfaulted lines.

    Ans. 0; $50 \mp j28.86$

6.27   The positive-, negative-, and zero-sequence reactances of a 15-MVA, 11-kV, three-phase, wye-connected generator are 11 percent, 8 percent, and 3 percent, respectively. The neutral of the generator is grounded, and the generator is excited to the rated voltage on open circuit. A line-to-ground fault occurs on phase a of the generator. Calculate the phase voltages and currents.

    Ans. 0 V; $4.922/254.74°$ kV; $4.922/-74.74°$ kV; $10,746/-18°$ A; 0 A; 0 A

6.28   A line-to-line fault occurs between phases b and c of the generator of Problem 6.27 while phase a remains open-circuited. Determine the phase voltages and currents.

    Ans. 5.35 kV; 2.67 kV; 2.67 kV; 0 A; 7169.6 A; -7169.6 A

6.29   The sequence reactances of a three-phase alternator are $X_1 = X_2 = 0.15$ pu and $X_0 = 0.05$ pu, based on the machine rating of 30 MVA and 13.2 kV. The alternator is connected to a transmission line having $X_1 = X_2 = 0.1$ pu and $X_0 = 0.4$ pu. A line-to-line fault occurs on the line at the receiving end. Draw the sequence networks for the faulted system, and calculate the line currents.

    Ans. Fig. 6-21; 0 A; $3788/180°$ A; $3788/0°$ A

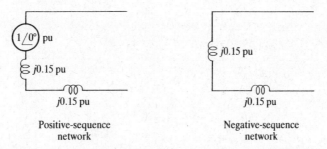

Fig. 6-21.

6.30   A double line-to-ground fault occurs at F in the system shown in Fig. 6-22. Draw the sequence networks for the system, and calculate the line current I_b.

    Ans. Fig. 6-23; $4469/123.8°$ A

6.31   Determine the subtransient currents (in amperes) in the two generators of Problem 6.16.

    Ans. 5720 A; 2860 A

6.32   A synchronous generator and a motor are rated at 30,000 kVA and 13.2 kV, and both have subtransient

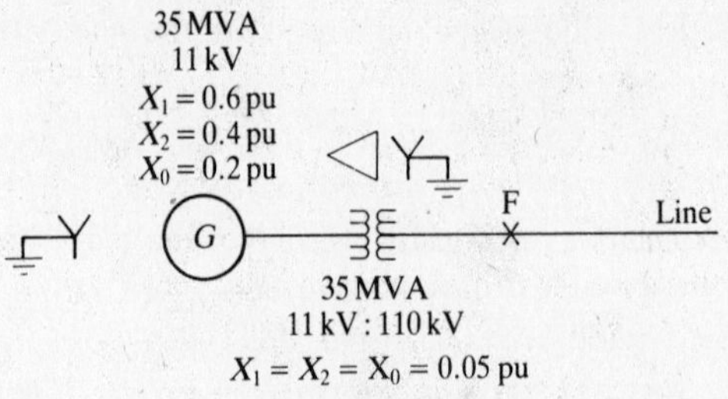

35 MVA
11 kV
$X_1 = 0.6$ pu
$X_2 = 0.4$ pu
$X_0 = 0.2$ pu

F   Line

35 MVA
11 kV : 110 kV
$X_1 = X_2 = X_0 = 0.05$ pu

Fig. 6-22.

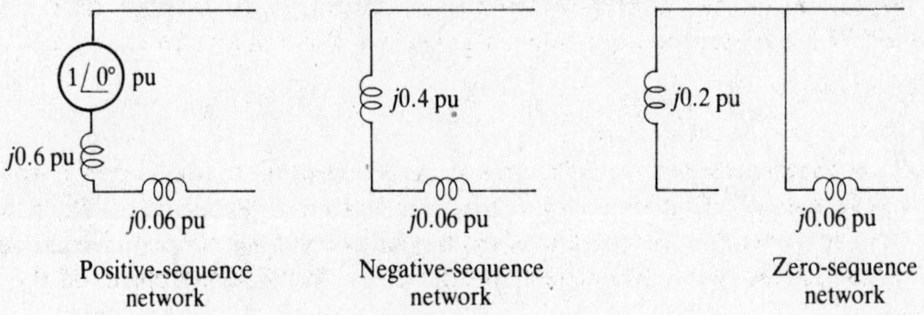

$1\underline{/0°}$ pu

$j0.6$ pu

$j0.06$ pu

Positive-sequence
network

$j0.4$ pu

$j0.06$ pu

Negative-sequence
network

$j0.2$ pu

$j0.06$ pu

Zero-sequence
network

Fig. 6-23.

reactances of 20 percent. The line connecting them has a reactance of 10 percent referred to the machine ratings. The motor is drawing 20,000 kW at 0.8 power factor leading and a terminal voltage of 12.8 kV when a symmetrical three-phase fault occurs at the motor terminals. Find the subtransient current in the generator. Draw an equivalent circuit to simulate this condition.

Ans. $3601\underline{/-75.7°}$ A; Fig. 6-24

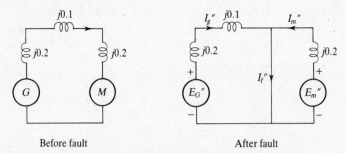

$j0.1$

$j0.2$ $j0.2$

G M

Before fault

I_g'' $j0.1$ I_m''

$j0.2$ $j0.2$

$+$ $+$

E_G'' I_f'' E_m''

$-$ $-$

After fault

Fig. 6-24.

6.33   What is the magnitude of the subtransient fault current in Problem 6.32?

Ans.   10.6 kA

6.34   For the system of Problem 6.32, choose 30 MVA and 13.2 kV as base values. Calculate the per-unit fault current by applying Thévenin's theorem.

Ans. $-j8.08$ pu

6.35   A three-phase short circuit occurs at F in the system of Fig. 6-25. The generator is loaded to 80 percent of its capacity at the time of the fault, and the receiving-end power factor is unity. Determine the rms per-unit current in one phase at F just after the occurrence of the fault.

Ans.   6.02 pu

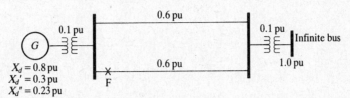

Fig. 6-25.

6.36 The per-unit reactances of a synchronous generator are $X_d = 1.1$, $X_d' = 0.24$, and $X_d'' = 0.15$. The generator is operating without load at 5 percent above rated voltage when a three-phase short circuit occurs at its terminals. What is the per-unit subtransient fault current? If the generator is rated at 500 MVA and 20 kV, determine the subtransient current in kiloamperes.

Ans. 7 pu; 101 kA

6.37 A synchronous generator having a subtransient reactance of 0.15 pu and operating at 5 percent above its rated voltage supplies a synchronous motor having a 0.20 pu subtransient reactance. The motor is connected to the generator by a transmission line and a transformer of total reactance 0.305 pu. A sudden three-phase short circuit occurs at the generator terminals. Determine the per-unit subtransient fault current.

Ans. $-j9.079$ pu

Chapter 7

General Methods for Network Calculations

In this chapter we develop general solution methods that are amenable to the computer solution of power system network problems. We begin from the basic network theorems.

7.1 SOURCE TRANSFORMATIONS

The voltage source of Fig. 7-1(a) may be transformed to the current source of Fig. 7-1(b) and vice versa, provided that

$$I_s = \frac{E_g}{Z_p} \tag{7.1}$$

and

$$Z_p = Z_g \tag{7.2}$$

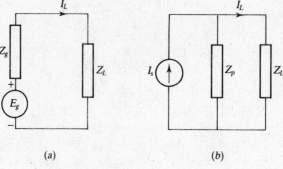

(a)   (b)

Fig. 7-1.

7.2 BUS ADMITTANCE MATRIX

The four-bus system that corresponds to the one-line diagram of Fig. 7-2(a) may be represented by the network of Fig. 7-2(b). In terms of the node voltages V_1, V_2, V_3 and V_4 and the given admittances, Kirchhoff's current law yields

$$I_1 = V_1 y_{10} + (V_1 - V_2)y_{12} + (V_1 - V_3)y_{13}$$
$$I_2 = V_2 y_{20} + (V_2 - V_1)y_{12} + (V_2 - V_3)y_{23} + (V_2 - V_4)y_{24}$$
$$I_3 = V_3 y_{30} + (V_3 - V_1)y_{13} + (V_3 - V_2)y_{23} + (V_3 - V_4)y_{34}$$
$$I_4 = V_4 y_{40} + (V_4 - V_2)y_{24} + (V_4 - V_3)y_{34}$$

Rearranging these equations and rewriting them in matrix form, we obtain

$$
\begin{bmatrix} I_1 \\ I_2 \\ I_3 \\ I_4 \end{bmatrix} =
\begin{bmatrix}
y_{10} + y_{12} + y_{13} & -y_{12} & -y_{13} & 0 \\
-y_{12} & y_{20} + y_{12} + y_{23} + y_{24} & -y_{23} & -y_{24} \\
-y_{13} & -y_{23} & y_{30} + y_{13} + y_{23} + y_{34} & -y_{34} \\
0 & -y_{24} & -y_{34} & y_{40} + y_{24} + y_{34}
\end{bmatrix}
\begin{bmatrix} V_1 \\ V_2 \\ V_3 \\ V_4 \end{bmatrix}
\tag{7.3}
$$

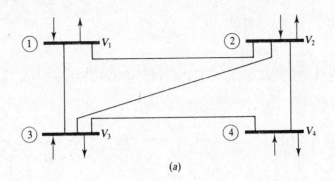

(a)

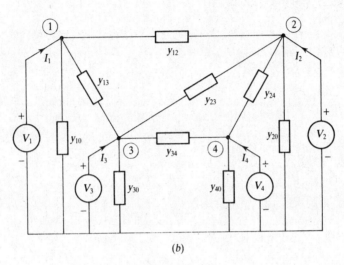

(b)

Fig. 7-2.

Equation (7.3) may be written as

$$\begin{bmatrix} I_1 \\ I_2 \\ I_3 \\ I_4 \end{bmatrix} = \begin{bmatrix} Y_{11} & Y_{12} & Y_{13} & Y_{14} \\ Y_{21} & Y_{22} & Y_{23} & Y_{24} \\ Y_{31} & Y_{32} & Y_{33} & Y_{34} \\ Y_{41} & Y_{42} & Y_{43} & Y_{44} \end{bmatrix} \begin{bmatrix} V_1 \\ V_2 \\ V_3 \\ V_4 \end{bmatrix}$$

(7.4)

where

$$Y_{11} = y_{10} + y_{12} + y_{13}$$

$$Y_{22} = y_{20} + y_{12} + y_{23} + y_{24}$$

$$Y_{33} = y_{30} + y_{13} + y_{23} + y_{34}$$

$$Y_{44} = y_{40} + y_{24} + y_{34}$$

$$Y_{12} = Y_{21} = -y_{12}$$

$$Y_{13} = Y_{31} = -y_{13}$$

$$Y_{14} = Y_{41} = -y_{14} = 0$$

$$Y_{23} = Y_{32} = -y_{23}$$

$$Y_{24} = Y_{42} = -y_{24}$$

$$Y_{34} = Y_{43} = -y_{34}$$

Each admittance Y_{ii} ($i = 1, 2, 3, 4$) is called the *self-admittance* (or *driving-point admittance*) of node i and is equal to the algebraic sum of all the admittances terminating on the node. Each off-diagonal term Y_{ik} ($i, k = 1, 2, 3, 4$) is called the *mutual admittance* (or *transfer admittance*) between nodes i and k and is equal to the negative of the sum of all the admittances connected directly between those nodes. Further, $Y_{ik} = Y_{ki}$.

For a general network with N nodes, therefore, Kirchhoff's current law in terms of node voltages may be written as

$$I = Y_{bus}V \tag{7.5}$$

where

$$Y_{bus} = \begin{bmatrix} Y_{11} & Y_{12} & \cdots & Y_{1N} \\ Y_{21} & Y_{22} & \cdots & Y_{2N} \\ \cdots\cdots\cdots\cdots\cdots\cdots\cdots \\ Y_{N1} & Y_{N2} & \cdots & Y_{NN} \end{bmatrix} \tag{7.6}$$

is called the *bus admittance matrix*, and V and I are the N-element *node voltage matrix* and *node current matrix*, respectively.

In (7.6), the first subscript on each Y indicates the node at which the current is being expressed, and the second subscript indicates the node whose voltage is responsible for a particular component of the current. Further, the admittances along the diagonal are the self-admittances, and the off-diagonal admittances are the mutual admittances. It follows from (7.5) and (7.6) that the current entering a node k is given by

$$I_k = \sum_{n=1}^{N} Y_{kn}V_n \tag{7.7}$$

For a large system, the matrices of (7.3) through (7.6) may be of correspondingly large order. In such a case, matrix operations can be more conveniently carried out when the matrices have been partitioned, or (loosely) subdivided. As an example, the matrix

$$A = \begin{bmatrix} a_{11} & a_{12} & \vdots & a_{13} \\ a_{21} & a_{22} & \vdots & a_{23} \\ \cdots\cdots\cdots\cdots\cdots\cdots\cdots \\ a_{31} & a_{32} & \vdots & a_{33} \end{bmatrix} \tag{7.8}$$

may be partitioned along the dashed lines into four submatrices such that

$$A = \begin{bmatrix} D & E \\ F & G \end{bmatrix} \tag{7.9}$$

where the submatrices are defined by

$$D = \begin{bmatrix} a_{11} & a_{12} \\ a_{21} & a_{22} \end{bmatrix} \qquad E = \begin{bmatrix} a_{13} \\ a_{23} \end{bmatrix} \tag{7.10}$$

$$F = \begin{bmatrix} a_{31} & a_{32} \end{bmatrix} \qquad G = a_{33}$$

To demonstrate matrix multiplication in terms of submatrices, let us assume that A of (7.8) is to be postmultiplied by a matrix

$$B = \begin{bmatrix} b_{11} \\ b_{21} \\ ----- \\ b_{31} \end{bmatrix} = \begin{bmatrix} H \\ J \end{bmatrix} \tag{7.11}$$

where
$$\mathbf{H} = \begin{bmatrix} b_{11} \\ b_{21} \end{bmatrix} \qquad \mathbf{J} = b_{31} \qquad\qquad (7.12)$$

(Note the correspondence between the vertical partitioning of $\mathbf{A}$ and the horizontal partitioning of $\mathbf{B}$.) The product is then written
$$\mathbf{C} = \mathbf{AB} = \begin{bmatrix} \mathbf{D} & \mathbf{E} \\ \mathbf{F} & \mathbf{G} \end{bmatrix} \begin{bmatrix} \mathbf{H} \\ \mathbf{J} \end{bmatrix} \qquad\qquad (7.13)$$

The submatrices are treated as matrix elements in the multiplication, so we obtain
$$\mathbf{C} = \begin{bmatrix} \mathbf{DH} + \mathbf{EJ} \\ \mathbf{FH} + \mathbf{GJ} \end{bmatrix} \qquad\qquad (7.14)$$

The product matrix $\mathbf{C}$ is finally obtained by performing the indicated multiplications and additions. For example, the bottom row of $\mathbf{C}$ turns out to be
$$\mathbf{FH} + \mathbf{GJ} = [a_{31} \quad a_{32}] \begin{bmatrix} b_{11} \\ b_{21} \end{bmatrix} + a_{33}b_{31}$$
$$= a_{31}b_{11} + a_{32}b_{21} + a_{33}b_{31} \qquad\qquad (7.15)$$

which is exactly what is found by multiplying the original matrices $\mathbf{A}$ and $\mathbf{B}$.

In solving (7.5), nodes at which no current enters or leaves may be eliminated. For this purpose, the matrices $\mathbf{I}$ and $\mathbf{V}$ must be rearranged so that elements pertaining to nodes to be eliminated are in the bottom rows of the matrices, and then partitioned accordingly. Then (7.5) can be rewritten as
$$\begin{bmatrix} \mathbf{I}_A \\ \mathbf{I}_X \end{bmatrix} = \begin{bmatrix} \mathbf{K} & \mathbf{L} \\ \mathbf{L}^T & \mathbf{M} \end{bmatrix} \begin{bmatrix} \mathbf{V}_A \\ \mathbf{V}_X \end{bmatrix} \qquad\qquad (7.16)$$

where $\mathbf{I}_X$ and $\mathbf{V}_X$ are the submatrices of currents and voltages, respectively, at the nodes to be eliminated. The submatrix $\mathbf{K}$ consists of admittances relating to the nodes to be retained; the submatrix $\mathbf{M}$ consists of admittances to the nodes to be eliminated; and the submatrix $\mathbf{L}$ and its transpose $\mathbf{L}^T$ consist of mutual admittances relating to both types of nodes. Solving for $\mathbf{I}_A$ from (7.16), we obtain
$$\mathbf{I}_A = [\mathbf{K} - \mathbf{L}\mathbf{M}^{-1}\mathbf{L}^T]\mathbf{V}_A$$
$$= \mathbf{Y}_{\text{bus}}\mathbf{V}_A \qquad\qquad (7.17)$$

where we now have the bus admittance matrix
$$\mathbf{Y}_{\text{bus}} = \mathbf{K} - \mathbf{L}\mathbf{M}^{-1}\mathbf{L}^T \qquad\qquad (7.18)$$

The inversion of $\mathbf{M}$ in (7.18) may become cumbersome, especially if $\mathbf{M}$ is a large matrix. In such a case, nodes may be eliminated one at a time, with the highest numbered node being eliminated each time. The elimination of that node transforms an $n \times n$ matrix into an $(n-1) \times (n-1)$ matrix. In addition, the remaining elements Y_{kj} of the original matrix must be modified to
$$Y_{kj(\text{new})} = Y_{kj(\text{orig})} - \frac{Y_{kn}Y_{nj}}{Y_{nn}} \qquad\qquad (7.19)$$

when the nth node is eliminated.

7.3 ELEMENTS OF $\mathbf{Y}_{\text{bus}}$

To find an element of (7.6), say Y_{22}, we write the second of the equations represented by (7.5) as
$$I_2 = Y_{21}V_1 + Y_{22}V_2 + Y_{23}V_3 + \cdots + Y_{2N}V_N \qquad\qquad (7.20)$$

from which we obtain

$$Y_{22} = \frac{I_2}{V_2}\bigg|_{V_1=V_3=\cdots=V_N=0} \tag{7.21}$$

Thus, the self-admittance of a given node is measured by shorting all other nodes and finding the ratio of the current injected at the node to the resulting voltage at that node. Similarly, we would find a mutual admittance, say Y_{21}, as

$$Y_{21} = \frac{I_2}{V_1}\bigg|_{V_2=V_3=\cdots=V_N=0} \tag{7.22}$$

7.4 BUS IMPEDANCE MATRIX

The bus impedance matrix is defined by

$$\mathbf{Z}_{\text{bus}} = \mathbf{Y}_{\text{bus}}^{-1} \tag{7.23}$$

Then (7.5) may be written as

$$\mathbf{V} = \mathbf{Z}_{\text{bus}}\mathbf{I} \tag{7.24}$$

where
$$\mathbf{Z}_{\text{bus}} = \begin{bmatrix} Z_{11} & Z_{12} & \cdots & Z_{1N} \\ Z_{21} & Z_{22} & \cdots & Z_{2N} \\ \cdots\cdots\cdots\cdots\cdots\cdots\cdots \\ Z_{N1} & Z_{N2} & \cdots & Z_{NN} \end{bmatrix} \tag{7.25}$$

7.5 ELEMENTS OF $\mathbf{Z}_{\text{bus}}$

The procedure for determining the elements of (7.25) is similar to that discussed in Section 7.4. Thus, the second of the equations represented by (7.24) may be written as

$$V_2 = Z_{21}I_1 + Z_{22}I_2 + Z_{23}I_3 + \cdots + Z_{2N}I_N \tag{7.26}$$

from which we obtain

$$Z_{22} = \frac{V_2}{I_2}\bigg|_{I_1=I_3=\cdots=I_N=0} \tag{7.27}$$

and

$$Z_{21} = \frac{V_2}{I_1}\bigg|_{I_2=I_3=\cdots=I_N=0} \tag{7.28}$$

Bus impedance and admittance matrices are used in power-flow studies (discussed in the next chapter) and in computer-aided fault calculations.

7.6 MODIFYING $\mathbf{Z}_{\text{bus}}$

The addition of a branch having an impedance Z_b to a system having an original bus impedance matrix $\mathbf{Z}_{\text{bus}}$ may take one of the following forms:

1. Inserting Z_b from the reference bus r to a new bus p (Fig. 7-3)

2. Inserting Z_b from an old bus k to a new bus p (Fig. 7-4)

3. Inserting Z_b from the reference bus r to an old bus k (Fig. 7-5)

4. Inserting Z_b between two old buses k and m (Fig. 7-6)

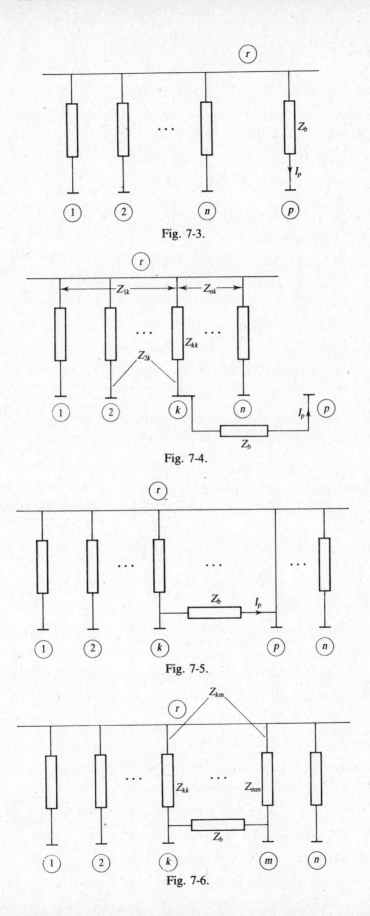

Fig. 7-3.

Fig. 7-4.

Fig. 7-5.

Fig. 7-6.

The new bus impedance matrices that result from the addition of Z_b are, respectively,

1. For Fig. 7-3:

$$
\mathbf{Z}_{\text{bus(new)}} = \left[
\begin{array}{cccc:c}
 & & & & 0 \\
 & \mathbf{Z}_{\text{bus}} & & & 0 \\
 & & & & \cdots \\
 & & & & 0 \\
\hdashline
0 & 0 & \cdots & 0 & Z_b
\end{array}
\right]
\tag{7.29}
$$

2. For Fig. 7-4:

$$
\mathbf{Z}_{\text{bus(new)}} = \left[
\begin{array}{cccc:c}
 & & & & Z_{1k} \\
 & \mathbf{Z}_{\text{bus}} & & & Z_{2k} \\
 & & & & \cdots \\
 & & & & Z_{Nk} \\
\hdashline
Z_{1k} & Z_{2k} & \cdots & Z_{Nk} & Z_{kk} + Z_b
\end{array}
\right]
\tag{7.30}
$$

3. For Fig. 7-5:

$$
\mathbf{Z}_{\text{bus(new)}} = \mathbf{Z}_{\text{bus}} - \frac{1}{Z_{kk} + Z_b}
\begin{bmatrix}
Z_{1k}^2 & Z_{1k}Z_{2k} & \cdots & Z_{1k}Z_{Nk} \\
Z_{2k}Z_{1k} & Z_{2k}^2 & \cdots & Z_{2k}Z_{Nk} \\
\cdots & \cdots & \cdots & \cdots \\
Z_{Nk}Z_{1k} & Z_{Nk}Z_{2k} & \cdots & Z_{Nk}^2
\end{bmatrix}
\tag{7.31}
$$

4. For Fig. 7-6:

$$
\mathbf{Z}_{\text{bus(new)}} = \mathbf{Z}_{\text{bus}} - \frac{1}{Z_b + Z_{kk} + Z_{mm} - 2Z_{km}}
$$

$$
\begin{bmatrix}
(Z_{1k} - Z_{1m})^2 & (Z_{1k} - Z_{1m})(Z_{2k} - Z_{2m}) & \cdots & (Z_{1k} - Z_{1m})(Z_{Nk} - Z_{Nm}) \\
(Z_{2k} - Z_{2m})(Z_{1k} - Z_{1m}) & (Z_{2k} - Z_{2m})^2 & \cdots & (Z_{2k} - Z_{2m})(Z_{Nk} - Z_{Nm}) \\
\cdots & \cdots & \cdots & \cdots \\
(Z_{Nk} - Z_{Nm})(Z_{1k} - Z_{1m}) & (Z_{Nk} - Z_{Nm})(Z_{2k} - Z_{2m}) & \cdots & (Z_{Nk} - Z_{Nm})^2
\end{bmatrix}
\tag{7.32}
$$

Solved Problems

7.1 The reactance diagram for a system is shown in Fig. 7-7(a). Use a source transformation to obtain the admittance diagram for the system. (All values are per-unit values.)

   The result of straightforward source transformation is shown in Fig. 7-7(b).

7.2 Obtain the bus admittance matrix for the network shown in Fig. 7-7(b).

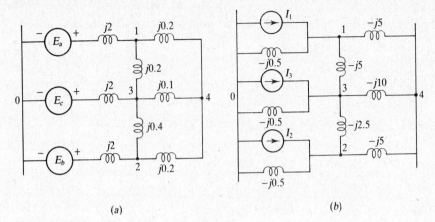

Fig. 7-7.

The various admittances are from the figure:

$$Y_{11} = -j0.5 - j5 - j5 = -j10.5$$
$$Y_{22} = -j0.5 - j2.5 - j5 = -j8.0$$
$$Y_{33} = -j0.5 - j5 - j10 - j2.5 = -j18.0$$
$$Y_{44} = -j5 - j10 - j5 = -j20.0$$
$$Y_{12} = Y_{21} = 0$$
$$Y_{13} = Y_{31} = j5.0$$
$$Y_{14} = Y_{41} = j5.0$$
$$Y_{23} = Y_{32} = j2.5$$
$$Y_{24} = Y_{42} = j5.0$$
$$Y_{23} = Y_{43} = j10.0$$

Hence the bus admittance matrix is

$$\mathbf{Y}_{bus} = \begin{bmatrix} -j10.5 & 0 & j5.0 & j5.0 \\ 0 & -j8.0 & j2.5 & j5.0 \\ j5.0 & j2.5 & -j18.0 & j10.0 \\ j5.0 & j5.0 & j10.0 & -j20.0 \end{bmatrix}$$

7.3 Draw an impedance diagram for the network shown in Fig. 7-8(a).

The impedance diagram is shown in Fig. 7-8(b), where

$$E_a = (2\underline{/0°})(-j1.0) = 2\underline{/-90°}\ \text{V}$$

and

$$E_b = (2\underline{/45°})(-j2.0) = 4\underline{/-45°}\ \text{V}$$

7.4 Obtain the bus admittance matrix for the network of Fig. 7-8(a). (All values are per-unit values.)

The elements of $\mathbf{Y}_{bus}$ are, from the figure,

$$Y_{11} = -j1.0 - j2.0 = -j3.0$$
$$Y_{22} = -j2.0 - j2.0 = -j4.0$$
$$Y_{12} = Y_{21} = 2.0$$

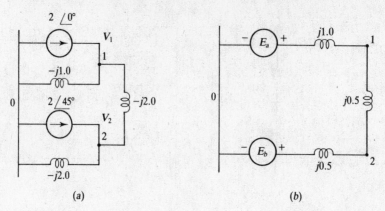

Fig. 7-8.

Thus, the bus admittance matrix is

$$\mathbf{Y}_{\text{bus}} = \begin{bmatrix} -j3.0 & j2.0 \\ j2.0 & -j4.0 \end{bmatrix}$$

7.5   By inverting* the bus admittance matrix for the network of Fig. 7-8(a), obtain the bus voltages V_1 and V_2.

From (7.5) we have

$$\begin{bmatrix} 2\underline{/0^\circ} \\ 2\underline{/45^\circ} \end{bmatrix} = \begin{bmatrix} -j3.0 & j2.0 \\ j2.0 & -j4.0 \end{bmatrix}\begin{bmatrix} V_1 \\ V_2 \end{bmatrix}$$

so

$$\begin{bmatrix} V_1 \\ V_2 \end{bmatrix} = \begin{bmatrix} -j3.0 & j2.0 \\ j2.0 & -j4.0 \end{bmatrix}^{-1}\begin{bmatrix} 2\underline{/0^\circ} \\ 2\underline{/45^\circ} \end{bmatrix} = \begin{bmatrix} j0.5 & j0.25 \\ j0.25 & j0.375 \end{bmatrix}\begin{bmatrix} 2\underline{/0^\circ} \\ 2\underline{/45^\circ} \end{bmatrix}$$

Then $V_1 = (j0.5)(2\underline{/0^\circ}) + (j0.25)(2\underline{/45^\circ}) = 1.848\underline{/67.5^\circ}\,\text{pu}$

and $V_2 = (j0.25)(2\underline{/0^\circ}) + (j0.375)(2\underline{/45^\circ}) = 1.158\underline{/117.2^\circ}\,\text{pu}$

7.6   Eliminate nodes 3 and 4 from the network of Fig. 7-7, using the procedure given by (7.16) through (7.18) to find the new $\mathbf{Y}_{\text{bus}}$.

From Problem 7.2 and (7.16) we have

$$\mathbf{Y}_{\text{bus}} = \begin{bmatrix} \mathbf{K} & \mathbf{L} \\ \mathbf{L}^T & \mathbf{M} \end{bmatrix} = \left[\begin{array}{cc:cc} -j10.5 & 0 & j5.0 & j5.0 \\ 0 & -j8.0 & j2.5 & j5.0 \\ \hdashline j5.0 & j2.5 & -j18.0 & j10.0 \\ j5.0 & j5.0 & j10.0 & -j20.0 \end{array} \right]$$

from which

$$\mathbf{M}^{-1} = -\frac{1}{260}\begin{bmatrix} -j20 & -j10 \\ -j10 & -j18 \end{bmatrix} = \begin{bmatrix} j0.0769 & j0.0385 \\ j0.0385 & j0.0692 \end{bmatrix}$$

Then $$\mathbf{LM}^{-1}\mathbf{L}^T = \begin{bmatrix} j5.0 & j5.0 \\ j2.5 & j5.0 \end{bmatrix}\begin{bmatrix} j0.0769 & j0.0385 \\ j0.0385 & j0.0692 \end{bmatrix}\begin{bmatrix} j5.0 & j2.5 \\ j5.0 & j5.0 \end{bmatrix} = \begin{bmatrix} -j5.575 & -j4.135 \\ -j4.135 & -j3.173 \end{bmatrix} \qquad (1)$$

---

* The matrix inversion procedure can be found in any text on matrix methods.

Hence, from (7.18),

$$\mathbf{Y}_{bus} = \begin{bmatrix} -j10.5 & 0 \\ 0 & -j8.0 \end{bmatrix} - \begin{bmatrix} -j5.575 & -j4.135 \\ -j4.135 & -j3.173 \end{bmatrix} = \begin{bmatrix} -j4.925 & j4.135 \\ j4.135 & -j4.827 \end{bmatrix} \text{pu} \qquad (2)$$

7.7   From the result of Problem 7.6, determine the per-unit impedance between buses (nodes) 1 and 2 of the circuit with nodes 3 and 4 eliminated. Also determine the admittance between buses 1 and 2 and the reference bus.

From (1) of Problem 7.6, the admittance between buses 1 and 2 is $-j4.135$. Hence,

$$\text{Per-unit impedance} = \frac{1}{-j4.135} = j0.2418 \text{ pu}$$

From (7.4), the admittance between bus 1 and the reference bus is $-j4.925 - (-j4.135) = -j0.79$ pu. Similarly, the admittance between bus 2 and the reference bus is $-j4.827 - (-j4.135) = -j0.692$ pu.

7.8   Draw an equivalent circuit, with voltage sources and with nodes 3 and 4 eliminated, for the network of Problem 7.6.

The required circuit, with values obtained in Problem 7.7 but with current sources, is shown in Fig. 7-9(a). The equivalent circuit with voltage sources is shown in Fig. 7-9(b).

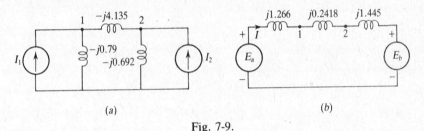

(a)   (b)

Fig. 7-9.

7.9   In Fig. 7-7(a), suppose that $E_a = 1.0\underline{/0°}$, $E_b = 1.2\underline{/-30°}$, and $E_c = 1.4\underline{/30°}$ per unit. Find, in complex form, the per-unit values of the current sources shown in Fig. 7-7(b).

The current sources are

$$I_1 = \frac{1.0\underline{/0°}}{j2.0} = 0.5\underline{/-90°} = -j0.5 \text{ pu}$$

$$I_2 = \frac{1.2\underline{/-30°}}{j2.0} = 0.6\underline{/-120°} = (-0.3 - j0.52) \text{ pu}$$

$$I_3 = \frac{1.4\underline{/30°}}{j2.0} = 0.7\underline{/-60°} = (0.35 - j0.606) \text{ pu}$$

7.10   If $E_a = 1.0\underline{/0°}$ and $E_b = 1.2\underline{/-30°}$ per unit in the circuit of Fig. 7-9, determine the complex power into (or out of) nodes 1 and 2.

From Fig. 7-9(b), the series impedance Z is $j(1.266 + 0.2418 + 1.445) = j2.9528$ pu. Hence

$$I = \frac{E_a - E_b}{Z} = \frac{1.0\underline{/0°} - 1.2\underline{/-30°}}{j2.9528}$$

$$= 0.2036\underline{/0°} \text{ pu}$$

Then   Power out of node $1 = (1.0\underline{/0°})(0.2036\underline{/0°}) = (0.2032 + j0)$ pu

and   Power into node $2 = (1.2\underline{/-30°})(0.2036\underline{/0°}) = (0.2116 - j0.122)$ pu

7.11 Determine the node voltages in the circuit of Fig. 7-9.

Using the current value determined in Problem 7.10, we have, at node 1,

$$V_1 = E_a - I(j1.266) = 1.0\underline{/0°} - (0.2036\underline{/0°})j1.266$$
$$= 10.32\underline{/-14.45°} \text{ pu}$$

Similarly, at node 2,

$$V_2 = E_a - I(j1.266 + j0.2418) = 1.0\underline{/0°} - (0.2036\underline{/0°})(j1.5078)$$
$$= 1.046\underline{/-17.06°} \text{ pu}$$

7.12 Eliminate nodes 3 and 4 of the network of Fig. 7-7, using (7.19) to find the new $\mathbf{Y}_{\text{bus}}$.

We begin with the bus admittance matrix of Problem 7.2:

$$\mathbf{Y}_{\text{bus}} = \begin{bmatrix} -j10.5 & 0 & j5.0 & j5.0 \\ 0 & -j8.0 & j2.5 & j5.0 \\ j5.0 & j2.5 & -j18.0 & j10.0 \\ j5.0 & j5.0 & j10.0 & -j20.0 \end{bmatrix}$$

We must remove the highest numbered node (here node 4) in each elimination step. We do so by modifying each of the remaining elements according to (7.19). With $n = 4$, for Y_{11} we have

$$Y_{11(\text{new})} = Y_{11(\text{orig})} - \frac{Y_{14}Y_{41}}{Y_{44}} = -j10.5 - \frac{(j5.0)(j5.0)}{-j20} = -j9.25$$

Similarly,

$$Y_{12(\text{new})} = 0 - \frac{(j5.0)(j5.0)}{-j20} = j1.25 = Y_{21(\text{new})}$$

$$Y_{13(\text{new})} = j5.0 - \frac{(j5)(j10)}{-j20} = j7.5 = Y_{31(\text{new})}$$

$$Y_{22(\text{new})} = -j8.0 - \frac{(j5)(j5)}{j20} = -j6.75$$

$$Y_{23(\text{new})} = j2.5 - \frac{(j10.0)(j5.0)}{-j20.0} = j5.0 = Y_{32(\text{new})}$$

$$Y_{33(\text{new})} = -j18.0 - \frac{(j10)(j10)}{j20} = -j13.0$$

Hence, the first elimination results in

$$\mathbf{Y}_{\text{bus(new)}} = \begin{bmatrix} -j9.25 & j1.25 & j7.5 \\ j1.25 & -j6.75 & j5.0 \\ j7.5 & j5.0 & -j13.0 \end{bmatrix}$$

The highest numbered node is now node 3; with $n = 3$, (7.19) yields

$$Y_{11(\text{new})} = -j9.25 - \frac{(j7.5)(j7.5)}{-j13.0} = -j4.923$$

And similarly for the remaining elements. Finally, with nodes 3 and 4 eliminated, we obtain

$$\mathbf{Y}_{\text{bus}} = \begin{bmatrix} -j4.923 & j4.134 \\ j4.134 & -j4.826 \end{bmatrix}$$

Note that this result agrees with that of Problem 7.6.

7.13 Find the bus impedance matrix of the network of Fig. 7-7(a).

From (7.23) we have

$$\mathbf{Z}_{bus} = [\mathbf{Y}_{bus}]^{-1} = \begin{bmatrix} -j10.5 & 0 & j5.0 & j5.0 \\ 0 & -j8.0 & j2.5 & j5.0 \\ j5.0 & j2.5 & -j18.0 & j10.0 \\ j5.0 & j5.0 & j10.0 & -j20.0 \end{bmatrix}^{-1}$$

$$= \begin{bmatrix} j0.724 & j0.620 & j0.656 & j0.644 \\ j0.620 & j0.738 & j0.642 & j0.660 \\ j0.656 & j0.642 & j0.702 & j0.676 \\ j0.664 & j0.660 & j0.676 & j0.719 \end{bmatrix}$$

7.14 For the emf's specified in Problem 7.9, find the voltage at node 4 in the network of Fig. 7-7.

With the current sources as computed in Problem 7.9, the node equation in matrix form is, from (7.5),

$$\begin{bmatrix} j0.5 \\ -0.3 - j0.52 \\ -0.35 - j0.606 \\ 0 \end{bmatrix} = \begin{bmatrix} -j10.5 & 0 & j5.0 & j5.0 \\ 0 & -j8.0 & j2.5 & j5.0 \\ j5.0 & j2.5 & -j18.0 & j10.0 \\ j5.0 & j5.0 & j10.0 & -j20.0 \end{bmatrix} \begin{bmatrix} V_1 \\ V_2 \\ V_3 \\ V_4 \end{bmatrix}$$

Using the result of Problem 7.13, where we found $\mathbf{Y}_{bus}^{-1}$, we have

$$\begin{bmatrix} V_1 \\ V_2 \\ V_3 \\ V_4 \end{bmatrix} = \begin{bmatrix} j0.724 & j0.620 & j0.656 & j0.664 \\ j0.620 & j0.738 & j0.642 & j0.660 \\ j0.656 & j0.642 & j0.702 & j0.676 \\ j0.664 & j0.660 & j0.676 & j0.719 \end{bmatrix} \begin{bmatrix} -j0.5 \\ -0.3 - j0.52 \\ -0.35 - j0.606 \\ 0 \end{bmatrix}$$

from which we find that

$$V_4 = (j0.664)(-j0.5) + (j0.66)(-0.3 - j0.52) + (j0.676)(-0.35 - j0.606)$$
$$= 1.085 - j0.435 = 1.169\underline{/-21.8°} \text{ pu}$$

7.15 A current of $-0.5\underline{/60°}$ pu is injected into node 4 of the network of Fig. 7-7. Find the resulting voltage at node 4, given the emf's specified in Problem 7.9.

We solve this problem by superposition. With the original emf's removed, and with Z_{44} as determined in Problem 7.13, the voltage at node 4 due to the injected current is

$$V_4' = I_4 Z_{44} = (-0.5\underline{/60°})(0.719\underline{/90°})$$
$$= -0.2514 + j0.2570 \text{ pu}$$

From Problem 7.14, the voltage at node 4 due to the emf's is

$$V_4'' = 1.085 - j0.435 \text{ pu}$$

Hence, the required voltage is the sum

$$V_4 = V_4' + V_4'' = -0.2514 + j0.2570 + 1.085 - j0.435$$
$$= 0.8336 - j0.178 = 0.851\underline{/-12.0°}$$

7.16 A capacitor having a reactance of 4.0 pu is connected from node 4 of the system of Fig. 7-7 to ground. With the numerical values specified in Fig. 7-7 and Problem 7.9, calculate the per-unit current through the capacitor.

From Problem 7.14, the Thévenin voltage at node 4 is $V_4 = 1.169\underline{/-21.8°}$ pu. And from Problem 7.13, the Thévenin impedance is $Z_{44} = j0.719$. Hence, the current through the capacitor is

$$I_c = \frac{1.169\underline{/-21.8°}}{j0.719 - j4.0} = 0.356\underline{/68.2°} \text{ pu}$$

Supplementary Problems

7.17   Determine $\mathbf{Z}_{bus}$ for the network shown in Fig. 7-10.

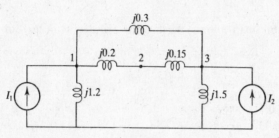

Fig. 7-10.

$$Ans. \quad \begin{bmatrix} j0.697 & j0.658 & j0.629 \\ j0.658 & j0.755 & j0.677 \\ j0.629 & j0.677 & j0.714 \end{bmatrix}$$

7.18   Find $\mathbf{Y}_{bus}$ for the network of Problem 7.17.

$$Ans. \quad \begin{bmatrix} -j9.15 & j4.98 & j3.33 \\ j4.98 & -j11.6 & j6.57 \\ j3.33 & j6.57 & -j10.6 \end{bmatrix} = \mathbf{Z}_{bus}^{-1}$$

7.19   Let $I_1 = 1.0\underline{/0°}$ pu and $I_2 = 1.2\underline{/30°}$ pu in the network of Fig. 7-10, and obtain an equivalent network having voltage sources only.

   Ans.   Fig. 7-11

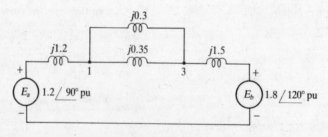

Fig. 7-11.

7.20   Find the complex power that enters (or leaves) node 1 of the network of Fig. 7-10.

   Ans. $(0.472 - j0.051)$ pu

7.21   Use a source transformation to obtain the impedance diagram for the network shown in Fig. 7-12.

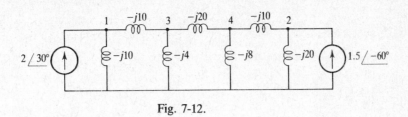

Fig. 7-12.

Ans.   Fig. 7-13

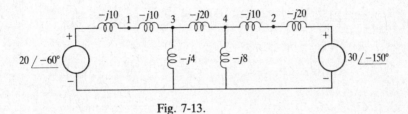

Fig. 7-13.

7.22   Obtain $\mathbf{Y}_{bus}$ for the network of Fig. 7-12.

Ans.
$$\begin{bmatrix} -j20 & 0 & -j10 & 0 \\ 0 & -j30 & 0 & -j10 \\ -j10 & 0 & -j34 & -j20 \\ 0 & -j10 & -j20 & -j38 \end{bmatrix}$$

7.23   Eliminate nodes 3 and 4 from the network of Fig. 7-12, using the procedure of (7.16) through (7.18) to obtain the resulting $\mathbf{Y}_{bus}$.

Ans.
$$\begin{bmatrix} -j15.74 & -j2.24 \\ -j2.24 & -j26.19 \end{bmatrix}$$

7.24   Draw an equivalent circuit, with voltage sources, for the network of Fig. 7-12 with nodes 3 and 4 eliminated.

Ans.   Fig. 7-14

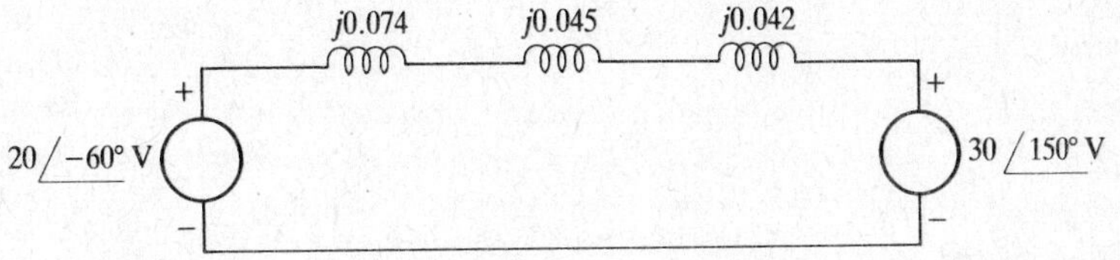

Fig. 7-14.

7.25   Repeat Problem 7.23, using (7.19) to find the new $\mathbf{Y}_{bus}$.

7.26   Find the voltage at node 3 under the conditions of Problem 7.15.

Ans. $1.25/-23.84°$ pu

7.27   If $\mathbf{Z}_{bus}$ is an $n \times n$ matrix, what are the orders of $\mathbf{Z}_{bus(new)}$ in (7.29) through (7.32)?

Ans. $(n + 1) \times (n + 1)$; $(n + 1) \times (n + 1)$; $n \times n$; $n \times n$

7.28   A two-bus system has $\mathbf{Z}_{\text{bus}} = \begin{bmatrix} j0.11565 & j0.04580 \\ j0.04580 & j0.13893 \end{bmatrix}$ pu. If an impedance $Z_b = j0.4$ pu is connected between buses 1 and 2, what is the new $\mathbf{Z}_{\text{bus}}$?

Ans. $\begin{bmatrix} j0.10698 & j0.05735 \\ j0.05735 & j0.12352 \end{bmatrix}$ pu

7.29   Find $\mathbf{Z}_{\text{bus}}$ for the system shown in Fig. 7-15. All impedances are per-unit values.

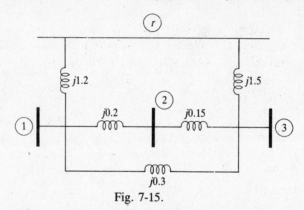

Fig. 7-15.

Ans. $\begin{bmatrix} j1.2 & j1.2 & j1.2 \\ j1.2 & j1.4 & j1.2 \\ j1.2 & j1.2 & j1.5 \end{bmatrix}$ pu

7.30   If an impedance of $j1.5$ pu is connected between bus r and bus 3 of Fig. 7-15, what is the new $\mathbf{Z}_{\text{bus}}$?

Ans. $\begin{bmatrix} j0.72 & j0.72 & j0.60 \\ j0.72 & j0.92 & j0.60 \\ j0.60 & j0.60 & j0.75 \end{bmatrix}$ pu

7.31   If an impedance of $j0.15$ pu is added between bus 2 and bus 3 of Fig. 7-15, what is the new bus impedance matrix?

Ans. $\begin{bmatrix} j0.6968 & j0.6581 & j0.6290 \\ j0.6581 & j0.7548 & j0.6774 \\ j0.6290 & j0.6774 & j0.7137 \end{bmatrix}$ pu

7.32   Impedance Z_{11} is the impedance that is measured between bus 1 and bus r of the system of Fig. 7-15 when $I_2 = I_3 = 0$. Evaluate Z_{11}. Verify that your result is consistent with that of Problem 7.31.

Ans. $j0.6968$ pu

7.33   A one-line diagram for a four-bus system is shown in Fig. 7-16. The line impedances are given in Table 7-1. Determine $\mathbf{Y}_{\text{bus}}$.

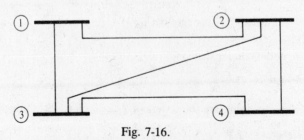

Fig. 7-16.

TABLE 7-1

Line (bus to bus)	R, pu	X, pu
1–2	0.05	0.15
1–3	0.10	0.30
2–3	0.15	0.45
2–4	0.10	0.30
3–4	0.05	0.15

$$Ans. \quad \begin{bmatrix} 3-j9 & -2+j6 & -1+j3 & 0 \\ -2+j6 & 3.666-j11 & -0.666+j2 & -1+j3 \\ -1+j3 & -0.666+j2 & 3.666-j11 & -2+j6 \\ 0 & -1+j3 & -2+j6 & 3-j9 \end{bmatrix} pu$$

Chapter 8

Power-Flow Studies

Power-flow studies, more commonly known as *load-flow studies,* are extremely important in evaluating the operation of power systems, controlling them, and planning for future expansions. A power-flow study yields mainly the real and reactive power and phasor voltage at each bus on the system, although much additional information is available from the computer printouts of typical power-flow studies.

The principles involved in power-flow studies are straightforward, but a study relating to a real power system can be carried out only with a digital computer. Then the required numerical computations are performed systematically by means of an iterative procedure; two of the more commonly used iterative numerical procedures are the Gauss-Seidel method and the Newton-Raphson method. Before considering these numerical methods, we illustrate the concept of power flow by obtaining explicit expressions for the power flow in a lossless short transmission line.

8.1 POWER FLOW IN A SHORT TRANSMISSION LINE

We assume the short transmission line shown in Fig. 8-1(a) has negligible resistance and a series reactance of jX ohms per phase. The per-phase sending-end and receiving-end voltages are V_S and V_R, respectively. We wish to determine the real and reactive power at the sending end and at the receiving end, given that V_S leads V_R by an angle δ.

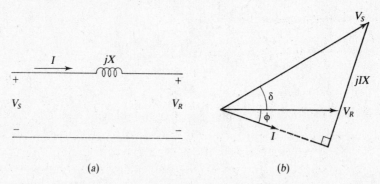

Fig. 8-1.

Complex power S, in voltamperes, is in general given by

$$S = P + jQ = VI^* \quad \text{VA} \tag{8.1}$$

where I^* is the complex conjugate of I. Thus, on a per-phase basis, at the sending end we have

$$S_S = P_S + jQ_S = V_S I^* \quad \text{VA} \tag{8.2}$$

From Fig. 8-1(a), I is given by

$$I = \frac{1}{jX}(V_S - V_R)$$

so

$$I^* = \frac{1}{-jX}(V_S^* - V_R^*) \tag{8.3}$$

112

Substituting (8.3) in (8.2) yields

$$S_S = \frac{V_S}{-jX}(V_S^* - V_R^*) \tag{8.4}$$

Now, from the phasor diagram of Fig. 8-1(b),

$$V_R = |V_R|\underline{/0^\circ} \quad \text{so} \quad V_R = V_R^*$$

and

$$V_S = |V_S|\underline{/\delta}$$

Hence (8.4) becomes

$$S_S = \frac{|V_S|^2 - |V_R||V_S|e^{j\delta}}{-jX}$$

$$= \frac{|V_S||V_R|}{X}\sin\delta + j\frac{1}{X}(|V_S|^2 - |V_S||V_R|\cos\delta)$$

(The latter equation requires some manipulation.) Finally, since $S_S = P_S + jQ_S$, we may write

$$P_S = \frac{1}{X}(|V_S||V_R|\sin\delta) \quad \text{W} \tag{8.5}$$

and

$$Q_S = \frac{1}{X}(|V_S|^2 - |V_S||V_R|\cos\delta) \quad \text{var} \tag{8.6}$$

Similarly, at the receiving end we have

$$S_R = P_R + jQ_R = V_R I^*$$

Proceeding as above, we finally obtain

$$P_R = \frac{1}{X}(|V_S||V_R|\sin\delta) \quad \text{W} \tag{8.7}$$

and

$$Q_R = \frac{1}{X}(|V_S||V_R|\cos\delta - |V_R|^2) \quad \text{var} \tag{8.8}$$

From this simple example, a number of significant conclusions may be derived. First, the transfer of real power depends only on the angle δ, which is known as the *power angle*, and not on the relative magnitudes of the sending-end and receiving-end voltages (unlike the case of a dc system). Moreover, the transmitted power varies approximately as the square of the voltage level. The maximum power transfer occurs when $\delta = 90^\circ$ and

$$P_{R(\max)} = P_{S(\max)} = \frac{|V_S||V_R|}{X} \tag{8.9}$$

Finally, from (8.6) and (8.8), it is clear that reactive power will flow in the direction of the lower voltage. If the system operates with $\delta \approx 0$, then the average reactive power flow over the line is given by

$$Q_{av} = \frac{1}{2}(Q_S + Q_R) = \frac{1}{2X}(|V_S|^2 - |V_R|^2) \quad \text{var} \tag{8.10}$$

This equation shows the strong dependence of the reactive power flow on the voltage difference.

To this point we have neglected the I^2R loss in the line. If we now assume that R is the resistance of the line per phase, then the line loss is given by

$$P_{line} = |I|^2R \quad \text{W} \tag{8.11}$$

From (8.2), we have

$$I^* = \frac{P + jQ}{V}$$

and

$$I = \frac{P - jQ}{V^*}$$

Thus,

$$II^* = |I|^2 = \frac{P^2 + Q^2}{|V|^2}$$

and (8.11) becomes

$$P_{\text{line}} = \frac{(P^2 + Q^2)R}{|V|^2} \quad \text{W} \tag{8.12}$$

indicating that both real and reactive power contribute to the line losses. Thus, it is important to reduce reactive power flow to reduce line losses.

8.2 AN ITERATIVE PROCEDURE

We were able to obtain an analytical expression for the power flow in our idealized case; however, in an actual power system, explicit analytical solutions are not forthcoming because of load fluctuations on the buses and because the receiving-end voltage may not be known. Then, numerical methods must be used to solve for unknown quantities—generally via an iterative procedure.

Figure 8-2 shows a two-bus system, with the real power represented by solid arrows and the reactive power by dashed arrows. The governing equations for the system are (on a per-phase basis)

$$S_2 = V_2 I^*$$
$$V_1 = V_2 + Z_\ell I$$

with the symbols as defined in Fig. 8-2. Solving for V_2 and eliminating I from these equations, we obtain

$$V_2 = V_1 - Z_\ell I = V_1 - Z_\ell \frac{S_2^*}{V_2^*} \tag{8.13}$$

To solve (8.13) iteratively, we would assume a value for V_2 and call it $V_2^{(0)}$. We would substitute this in the right-hand side of (8.13) and solve for V_2, calling the new value of V_2, obtained in this first iteration, $V_2^{(1)}$. We would then substitute $(V_2^{(1)})^*$ in the right-hand side of (8.13) and obtain a new value $V_2^{(2)}$. This procedure would be repeated until convergence to the desired precision was achieved. The iterative process we would use is thus given by the general equation, or *algorithm*,

$$V_2^{(k)} = V_1 - \frac{Z_\ell S_2^*}{(V_2^{(k-1)})^*} \tag{8.14}$$

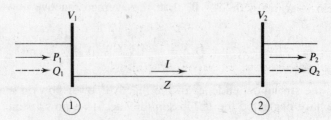

Fig. 8-2.

8.3 THE POWER-FLOW EQUATIONS

As noted in the last chapter, the bus admittance matrix is useful in a systematic approach to the solution of power-flow problems. Before discussing this approach, we need to define the following special buses:

1. A *load bus* is a bus for which the active and reactive powers P and Q are known, and $|V|$ and δ are to be found.

2. A *generator bus* is a bus for which the magnitude of the generated voltage $|V|$ and the corresponding generated power P are known, and Q and δ are to be obtained.

3. A *swing bus* (or *slack bus*) is a generator bus at which $|V|$ and δ are specified, and P and Q are to be determined. For convenience, we choose $V\underline{/\delta} = 1\underline{/0°}$ per unit.

From (7.5), we may write the kth (of N) nodal current as

$$I_k = \sum_{n=1}^{N} Y_{kn}V_n \tag{8.15}$$

which may also be written as

$$I_k = Y_{kk}V_k + \sum_{\substack{n=1 \\ n \neq k}}^{N} Y_{kn}V_n \tag{8.16}$$

Solving for V_k yields

$$V_k = \frac{I_k}{Y_{kk}} - \frac{1}{Y_{kk}} \sum_{\substack{n=1 \\ n \neq k}}^{N} Y_{kn}V_n \tag{8.17}$$

Now, since

$$V_k^* I_k = P_k - jQ_k \tag{8.18}$$

we have

$$I_k = \frac{P_k - jQ_k}{V_k^*} \tag{8.19}$$

Finally, (8.17) and (8.19) give, for N nodes,

$$V_k = \frac{1}{Y_{kk}} \left(\frac{P_k - jQ_k}{V_k^*} - \sum_{\substack{n=1 \\ n \neq k}}^{N} Y_{kn}V_n \right) \quad \text{for } k = 1, 2, \ldots, N \tag{8.20}$$

This set of N equations constitutes the power-flow equations.

8.4 THE GAUSS AND GAUSS–SEIDEL METHODS

The Gauss and Gauss–Seidel methods are iterative procedures for solving simultaneous (nonlinear) equations. We illustrate the Gauss method with the following example.

EXAMPLE: Solve for x and y in the system

$$y - 3x + 1.9 = 0$$
$$y + x^2 - 1.8 = 0$$

To solve with the Gauss method, we rewrite the given equations as

$$x = \frac{y}{3} + 0.633 \tag{1}$$

$$y = 1.8 - x^2 \tag{2}$$

We now make an initial guess of $x_0 = 1$ and $y_0 = 1$, update x with (1), and update y with (2). That is, we compute

$$x_1 = \frac{y_0}{3} + 0.633 = \frac{1}{3} + 0.633 = 0.9663 \tag{3}$$

and

$$y_1 = 1.8 - x_0^2 = 1.8 - 1 = 0.8 \tag{4}$$

In succeeding iterations we compute, more generally,

$$x_{n+1} = \frac{y_n}{3} + 0.633 \tag{5}$$

and

$$y_{n+1} = 1.8 - x_n^2 \tag{6}$$

After several iterations, we obtain $x = 0.938$ and $y = 0.917$. A few more iterations would bring us very close to the exact results: $x = 0.93926$ and $y = 0.9178$. However, it must be pointed out that an "uneducated guess" of the initial values (such as $x_0 = y_0 = 100$) would have caused the solution to diverge.

If we were to use the Gauss–Seidel method in the above example, we would still use (5) to compute x_{n+1}, but we would then use the just-computed x_{n+1} to find y_{n+1}. Instead of (5) and (6), the algorithm for the Gauss–Seidel method would be

$$x_{n+1} = \frac{y_n}{3} + 0.633$$

$$y_{n+1} = 1.8 - x_{n+1}^2$$

Extrapolating the above results, we find that the Gauss–Seidel algorithm for the power-flow equations (8.20) is

$$V_k^{(i+1)} = \frac{1}{Y_{kk}} \left[\frac{P_k - jQ_k}{(V_k^{(i)})^*} - \sum_{\substack{n=1 \\ n \neq k}}^{N} Y_{kn} V_n^{(i)} \right] \qquad \text{for } k = 2, 3, \ldots, N \tag{8.21}$$

Notice that V_1 in (8.21) is specified, so we begin the computations with bus 2.

8.5 THE NEWTON–RAPHSON METHOD

Consider two functions of two variables x_1 and x_2, such that

$$f(x_1, x_2) = C_1 \tag{8.22}$$

$$f_2(x_1, x_2) = C_2 \tag{8.23}$$

where C_1 and C_2 are constants. Let $x_1^{(0)}$ and $x_2^{(0)}$ be initial estimates of solutions to (8.22) and (8.23), and let $\Delta x_1^{(0)}$ and $\Delta x_2^{(0)}$ be the values by which the initial estimates differ from the correct solutions. That is,

$$f_1(x_1^{(0)} + \Delta x_1^{(0)}, x_2^{(0)} + \Delta x_2^{(0)}) = C_1 \tag{8.24}$$

$$f_2(x_1^{(0)} + \Delta x_1^{(0)}, x_2^{(0)} + \Delta x_2^{(0)}) = C_2 \tag{8.25}$$

Expanding the left-hand side of each of these equations in a Taylor's series, we obtain

$$f_1(x_1^{(0)}, x_2^{(0)}) + \Delta x_1^{(0)} \left. \frac{\partial f_1}{\partial x_1} \right|_{x_1^{(0)}} + \Delta x_2^{(0)} \left. \frac{\partial f_1}{\partial x_2} \right|_{x_2^{(0)}} + \cdots = C_1 \tag{8.26}$$

$$f_2(x_1^{(0)}, x_2^{(0)}) + \Delta x_1^{(0)} \left. \frac{\partial f_2}{\partial x_1} \right|_{x_1^{(0)}} + \Delta x_2^{(0)} \left. \frac{\partial f_2}{\partial x_2} \right|_{x_2^{(0)}} + \cdots = C_2 \tag{8.27}$$

Neglecting derivatives of order greater than one and writing the result in matrix form yields

$$\begin{bmatrix} C_1 - f_1(x_1^{(0)}, x_2^{(0)}) \\ C_2 - f_2(x_1^{(0)}, x_2^{(0)}) \end{bmatrix} = \begin{bmatrix} \dfrac{\partial f_1}{\partial x_1} & \dfrac{\partial f_1}{\partial x_2} \\ \dfrac{\partial f_2}{\partial x_1} & \dfrac{\partial f_2}{\partial x_2} \end{bmatrix}_{x_1^{(0)},\, x_2^{(0)}} \begin{bmatrix} \Delta x_1^{(0)} \\ \Delta x_2^{(0)} \end{bmatrix} \tag{8.28}$$

where the derivatives are evaluated at $x_1^{(0)}$ and $x_2^{(0)}$. Equation (8.28) may be abbreviated as

$$\begin{bmatrix} \Delta C_1^{(0)} \\ \Delta C_2^{(0)} \end{bmatrix} = \mathbf{J}^{(0)} \begin{bmatrix} \Delta x_1^{(0)} \\ \Delta x_2^{(0)} \end{bmatrix} \tag{8.29}$$

where the matrix $\mathbf{J}^{(0)}$ is called the *jacobian* (of the initial estimates), and $\Delta C_1^{(0)}$ and $\Delta C_2^{(0)}$ are the differences specified on the left side of (8.28).

Solution of the matrix equation (8.29) gives $\Delta x_1^{(0)}$ and $\Delta x_2^{(0)}$. Then a better estimate of the solutions is

$$x_1^{(1)} = x_1^{(0)} + \Delta x_1^{(0)} \tag{8.30}$$

$$x_2^{(1)} = x_2^{(0)} + \Delta x_2^{(0)} \tag{8.31}$$

Repeating the process with these values gives a still better estimate, and the iterations are continued until Δx_1 and Δx_2 become smaller than a predetermined value.

To apply the Newton–Raphson method to a power-flow problem, for the kth bus we let

$$V_k = |V_k| \,\underline{/\delta_k} \qquad V_n = |V_n| \,\underline{/\delta_n} \qquad Y_{kn} = |Y_{kn}| \,\underline{/\theta_{kn}}$$

Then, from (8.15) and (8.19),

$$P_k - jQ_k = \sum_{n=1}^{N} |V_k V_n Y_{kn}| \,\underline{/\theta_{kn} + \delta_n - \delta_k} \tag{8.32}$$

so that

$$P_k = \sum_{n=1}^{N} |V_k V_n Y_{kn}| \cos(\theta_{kn} + \delta_n - \delta_k) \tag{8.33}$$

and

$$Q_k = \sum_{n=1}^{N} |V_k V_n Y_{kn}| \sin(\theta_{kn} + \delta_n - \delta_k) \tag{8.34}$$

Having P and Q specified for every bus (except a swing bus) corresponds to knowing C_1 and C_2 in (8.28). We first estimate V and δ for each bus except the swing bus, for which they are known. We then substitute these estimated values, which correspond to the estimated values for x_1 and x_2, in (8.33) and (8.34) to calculate P's and Q's that correspond to $f_1(x_1^{(0)}, x_2^{(0)})$ and $f_2(x_1^{(0)}, x_2^{(0)})$. Next we compute

$$\Delta P_k^{(0)} = P_{ks} - P_{kc}^{(0)} \tag{8.35}$$

$$\Delta Q_k^{(0)} = Q_{ks} - Q_{kc}^{(0)} \tag{8.36}$$

where the subscripts s and c mean, respectively, specified and calculated values. These correspond to the values on the left side of (8.29).

Corresponding to (*8.28*) and (*8.29*), the matrix equation for a three-bus system (with bus 1 as the swing bus and hence omitted) is

$$
\begin{bmatrix} \Delta P_2^{(0)} \\ \Delta P_3^{(0)} \\ \Delta Q_2^{(0)} \\ \Delta Q_3^{(0)} \end{bmatrix} = \begin{bmatrix} \dfrac{\partial P_2}{\partial \delta_2} & \dfrac{\partial P_2}{\partial \delta_3} & \dfrac{\partial P_2}{\partial |V_2|} & \dfrac{\partial P_2}{\partial |V_3|} \\[2mm] \dfrac{\partial P_3}{\partial \delta_2} & \dfrac{\partial P_3}{\partial \delta_3} & \dfrac{\partial P_3}{\partial |V_2|} & \dfrac{\partial P_3}{\partial |V_3|} \\[2mm] \dfrac{\partial Q_2}{\partial \delta_2} & \dfrac{\partial Q_2}{\partial \delta_3} & \dfrac{\partial Q_2}{\partial |V_2|} & \dfrac{\partial Q_2}{\partial |V_3|} \\[2mm] \dfrac{\partial Q_3}{\partial \delta_2} & \dfrac{\partial Q_3}{\partial \delta_3} & \dfrac{\partial Q_3}{\partial |V_2|} & \dfrac{\partial Q_3}{\partial |V_3|} \end{bmatrix}_{\substack{\Delta \delta_2^{(0)},\, \Delta \delta_3^{(0)} \\ \Delta V_2^{(0)},\, \Delta V_3^{(0)}}} \begin{bmatrix} \Delta \delta_2^{(0)} \\ \Delta \delta_3^{(0)} \\ \Delta |V_2^{(0)}| \\ \Delta |V_3^{(0)}| \end{bmatrix} \qquad (8.37)
$$

Equation (*8.37*) is solved by inverting the jacobian. The values determined for $\Delta \delta_k^{(0)}$ and $\Delta V_k^{(0)}$ are added to the previous estimates of V and δ to obtain new estimates with which to start the next iteration. The process is repeated until the values in either column matrix are as small as desired.

8.6 BUS VOLTAGE SPECIFICATION AND REGULATION

In Sections 8.2 and 8.3 we indicated that bus voltages are specified in some power-flow studies. In certain cases the real power on each of the various buses is also specified. The corresponding reactive power is then determined as needed to maintain each bus voltage. Stated otherwise, we investigate the effect of a particular bus voltage on the reactive power supplied by a given generator to the system. If we represent the system with its Thévenin equivalent and connect the system to a generator as shown in Fig. 8-3, then the corresponding phasor diagram resembles those shown in Fig. 8-4 for leading, lagging, and unity power factors. Notice, in the three diagrams of Fig. 8-4, that for a constant power delivered by the generator, the component of I in phase with E_{Th} must be constant. It follows from Fig. 8-4 that with constant power input to the bus, larger magnitudes of bus voltage V_1 require larger $|E_g|$, and the larger $|E_g|$ is obtained by increasing the excitation of the generator. Increasing the bus voltage by increasing $|E_g|$ causes the current to become more lagging. Thus, increasing the voltage specified at a generator bus means that the generator feeding the bus will increase its output of reactive power to the bus. From the standpoint of the operation of the system, we are controlling bus voltage and Q generation by adjusting the generator excitation.

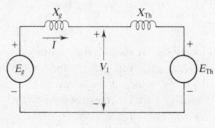

Fig. 8-3.

Another method of controlling bus voltage is by installing shunt capacitor banks at the buses at both the transmission and distribution voltage levels along transmission lines or at substations and loads. Each capacitor bank supplies reactive power at the point at which it is placed. Hence it reduces the line current necessary to supply reactive power to the load and reduces the voltage drop in the

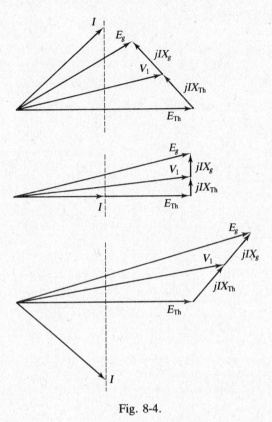

Fig. 8-4.

line via power-factor improvement. Since the capacitor banks lower the reactive power required from the generators, more real power output is available.

If a capacitor bank is installed at a particular node, the node (or bus) voltage can be determined from the Thévenin equivalent of the system, shown in Fig. 8-5, and the corresponding phasor diagram, shown in Fig. 8-6. The increase in V_1 due to the capacitor bank is approximately equal to $|I_C| X_{Th}$ if E_{Th} remains constant.

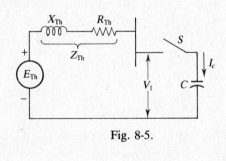

Fig. 8-5.

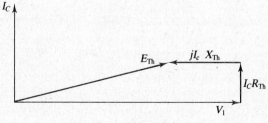

Fig. 8-6.

Solved Problems

8.1   For the system shown in Fig. 8-7,* it is desired that $|V_1| = |V_2| = 1$ pu. The loads, as shown, are $S_1 = 6 + j10$ pu and $S_2 = 14 + j8$ pu. The line impedance is $j0.05$ pu. If the real power input at each bus is 10 pu, calculate the power and the power factors at the two ends.

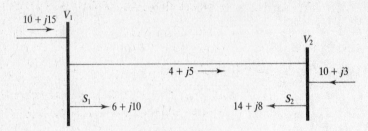

Fig. 8-7.

Let $V_2 = 1\underline{/0°}$ and $V_1 = 1\underline{/\delta}$. Then

$$P_1 = P_2 = \frac{|V_1||V_2|}{X}\sin\delta \quad \text{or} \quad 10 = \frac{1 \times 1}{0.05}\sin\delta$$

from which $\delta = 30°$ and $V_1 = 1\underline{/30°}$. The reactive power is given by

$$Q_1 = \frac{|V_1|^2}{X} - \frac{|V_1||V_2|}{X}\cos\delta = \frac{1}{0.05} - \frac{1}{0.05}\cos 30° = 2.68\text{ pu} = -Q_2$$

so $Q_{\text{line}} = Q_1 - Q_2 = 2Q_1 = 5.36$ pu

Thus we have

$$\text{Load on bus 1} = (6 + j10) + (10 + j5.36) = 16 + j15.36\text{ pu}$$

$$\text{Power factor at bus 1} = 0.72\text{ lagging}$$

$$\text{Load on bus 2} = (14 + j8) - (10 - j5.36) = 24 + j13.36\text{ pu}$$

$$\text{Power factor at bus 2} = 0.87\text{ lagging}$$

8.2   In Fig. 8-2, let $V_1 = 1\underline{/0°}$, $Z_\ell = 0.05 + j0.02$, and $P_2 + jQ_2 = 1.0 + j0.6$ (all per unit). Determine V_2 and $P_1 + jQ_1$.

Based on the given numerical values, we make the initial assumption $V_2 = 1\underline{/0°}$ and use (8.14) iteratively to obtain the following values:

Iteration	V_2, pu
0	$1.0 + j0$
1	$0.962 - j0.05$
2	$0.9630 - j0.054$
3	$0.9635 - j0.054$
4	$0.9635 - j0.054$

(Note that convergence is achieved in just four iterations. Different data, such as a greater load, might require more iterations for convergence to the solution. Or, convergence may not be achieved at all if a solution does not exist or if the starting point of the iteration process is not appropriate.)

---

* In Figs. 8-7 to 8-12, complex numbers denote per-unit apparent power.

Now, with $V_2 = 0.9635 - j0.054$ pu, we have

$$I = \frac{S_2^*}{V_2^*} = \frac{1.0 - j0.6}{0.9635 + j0.054} = 1.208\underline{/-27.75°} \text{ pu}$$

Then $I^* = 1.208\underline{/27.75°}$, and since $V_1 = 1\underline{/0°}$, we have

$$P_1 + jQ_1 = S_1 = V_1 I^* = (1\underline{/0°})(1.208\underline{/27.75°})$$
$$= 1.069 + j0.5625 \text{ pu}$$

8.3 For the system of Problem 8.2, it is desired to have $|V_1| = |V_2| = 1.0$ pu by supplying reactive power at bus 2. Determine the reactive power that must be supplied.

From (8.1) we obtain

$$I = \frac{S_2^* + jQ_2'}{V_2^*}$$

which, when substituted in (8.13), yields

$$V_1 = V_2 + \frac{Z_\ell}{V_2^*}(S_2^* + jQ_2') \tag{1}$$

Where Q_2' represents the added reactive power at bus 2. We now substitute in (1)

$$|V_1| = 1 \qquad V_2 = 1\underline{/0°} \qquad Z_\ell = 0.05 + j0.02 \qquad S_2^* = 1 - j0.6$$

and obtain, as the absolute value of the right-hand side,

$$1 = |1 + (0.05 + j0.02)[1 + j(Q_2' - 0.06)]|$$

Hence, $Q_2' = 4.02$ pu

8.4 Two buses are interconnected by a transmission line of impedance $(0.3 + j1.2)$ pu. The voltage on one bus is $1\underline{/0°}$, and the load on the other bus is $(1 + j0.4)$ pu. Determine the per-unit voltage on this second bus. Also calculate the per-unit real and reactive power on the first bus.

Equation (8.14) gives the following values for V_2:

Iteration	V_2, pu
0	$1 + j0$
1	$0.922 - j0.108$
2	$0.903 - j0.106$
3	$0.9 - j0.108$
4	$0.9 - j0.108$

Then $V_2 = 0.9 - j0.108$ pu and

$$I = \frac{S_2^*}{V_2^*} = \frac{1 - j0.4}{0.9 + j0.108} = 1.188\underline{/-28.6°} \text{ pu}$$

so that $I^* = 1.188\underline{/28.6°}$. This gives us

$$S_1 = P_1 + jQ_1 = V_1 I_1^* = (1\underline{/0})(1.188\underline{/28.6°})$$
$$= 1.188\underline{/28.6°} = 1.043 + j0.569$$

Hence, $P_1 = 1.043$ pu and $Q_1 = 0.569$ pu.

8.5 The voltages on the two buses of Problem 8.4 are to be made equal in magnitude by supplying reactive power at the second bus. How much reactive power must be supplied?

We have $|V_1| = 1$, $V_2 = 1\underline{/0°}$ (desired), $Z_e = 0.03 + j0.12$, and $S_2^* = (1 - j0.4)$. Then (1) of Problem 8.3 yields

$$1 = |1 + (0.03 + j0.12)[1 + j(Q_2 - 0.4)]|$$

Hence, $Q_2 = 0.259$ pu.

8.6 The per-unit impedance of a short transmission line is $j0.06$. The per-unit load on the line is $(1 + j0.6)$ pu at a receiving-end voltage of $1\underline{/0°}$ pu. Calculate the average reactive power flow over the line.

The sending-end voltage is

$$V_S = V_R + IZ = 1\underline{/0°} + (1 + j0.6)(j0.06) = 0.9658\underline{/3.56°} \text{ pu}$$

Thus, from (8.10),

$$Q_{av} = \frac{1}{2(0.06)}(0.9658^2 - 1^2) = -56 \text{ pu}$$

8.7 Solve the following equation by the Gauss–Seidel method: $x^2 - 6x + 2 = 0$.

We solve the given equation for x, obtaining

$$x = \tfrac{1}{6}x^2 + \tfrac{1}{3} = F(x)$$

and use the initial estimate $x^{(0)} = 1$. Then, in succeeding iterations we obtain

Iteration 1: $x^{(1)} = F(1) = \tfrac{1}{6} + \tfrac{1}{3} = 0.5$

Iteration 2: $x^{(2)} = F(0.5) = \tfrac{1}{6}(0.5)^2 + \tfrac{1}{3} = 0.375$

Iteration 3: $x^{(3)} = F(0.375) = \tfrac{1}{6}(0.375)^2 + \tfrac{1}{3} = 0.3568$

Iteration 4: $x^{(4)} = F(0.3568) = \tfrac{1}{6}(0.3568)^2 + \tfrac{1}{3} = 0.3545$

We may now stop since $|x^{(n+1)}| - |x^{(n)}| < \epsilon = 0.0023$; which seems to be sufficiently small. The quadratic formula gives this root as $x = 0.35425$ to five places.

8.8 For the two-bus system of Fig. 8-8, with the data as shown and with $Y_{11} = Y_{22} = 1.6\underline{/-80°}$ pu and $Y_{21} = Y_{12} = 1.9\underline{/100°}$ pu, determine the per-unit voltage at bus 2 by the Gauss–Seidel method.

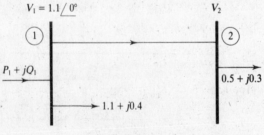

Fig. 8-8.

The power into the two buses is

$$S_1 = (P_1 - 1.1) + j(Q_1 - 0.4) \text{ pu}$$
$$S_2 = -0.5 - j0.3 \text{ pu}$$

From (8.21), we have the Gauss–Seidel algorithm

$$V_2^{(i+1)} = \frac{1}{Y_{22}}\left[\frac{P_2 - jQ_2}{(V_2^{(i)})^*} - Y_{21}V_1\right] \tag{1}$$

With the given numerical values, (1) becomes

$$V_2^{(i+1)} = \frac{1}{1.6\underline{/-80°}}\left[\frac{0.583\underline{/149°}}{(V_2^{(i)})^*} - (1.9\underline{/100°})(1.1\underline{/0°})\right]$$

$$= (0.625\underline{/80°})\left[\frac{0.583\underline{/149°}}{(V_2^{(i)})^*} - 2.09\underline{/100°}\right] \tag{2}$$

To begin the iterations, we let $V_2^{(0)} = 1.0\underline{/-10°}$ pu. Then (2) yields

$$V_2^{(1)} = (0.625\underline{/80°})[(0.583\underline{/139°}) - (2.09\underline{/100°})]$$

$$= 1.048\underline{/-12.6°}\text{ pu}$$

The next iteration yields

$$V_2^{(2)} = (0.625\underline{/80°})\left(\frac{0.583\underline{/149°}}{1.048\underline{/12.6°}} - 2.09\underline{/100°}\right)$$

$$= 1.047\underline{/-8.6°}\text{ pu}$$

Further iteration is unnecessary.

8.9   Compute the power on the swing bus of the network of Fig. 8-8.

From Problem 8.8 we have the per-unit values $V_1 = 1.1\underline{/0°}$, $V_2 = 1.047\underline{/-8.6°}$, $Y_{11} = 1.6\underline{/-80°}$, and $Y_{12} = 1.9\underline{/100°}$. Substituting these values in (8.32) with $k = 1$ yields the complex power for bus 1:

$$P_1 - jQ_1 = (1.1 \times 1.1 \times 1.6)\underline{/-80°} + (1.1 \times 1.047 \times 1.9)\underline{/100° - 8.6°}$$

$$= 0.3209 + j0.2816\text{ pu}$$

8.10   For the system shown in Fig. 8-9, the bus admittance matrix is

$$\mathbf{Y}_{bus} = \begin{bmatrix} 3 - j9 & -2 + j6 & -1 + j3 & 0 \\ -2 + j6 & 3.666 - j11 & -0.666 + j2 & -1 + j3 \\ -1 + j3 & -0.666 + j2 & 3.666 - j11 & -2 + j6 \\ 0 & -1 + j3 & -2 + j6 & 3 - j9 \end{bmatrix}\text{pu}$$

With the complex power on buses 2, 3, and 4 as shown in the figure determine the value for V_2 that is produced by the first iteration of the Gauss–Seidel procedure.

Let $V_2^{(0)} = V_3^{(0)} = V_4^{(0)} = 1.0\underline{/0°}$ pu. Then, from (8.21),

$$V_2^{(1)} = \frac{1}{Y_{22}}\left[\frac{P_2 - jQ_2}{(V_2^{(0)})^*} - Y_{21}V_1 - Y_{23}V_3^{(0)} - Y_{24}V_4^{(0)}\right]$$

$$= \frac{1}{Y_{22}}\left[\frac{0.5 + j0.2}{1 - j0} - 1.04(-2 + j6) - (-0.666 + j2) - (-1 + j3)\right]$$

$$= \frac{4.246 - j11.04}{3.666 - j11} = 1.019 + j0.046$$

$$= 1.02\underline{/2.58°}\text{ pu}$$

8.11   Determine the value for V_2 of Problem 8.10 as produced by the second iteration of the Gauss–Seidel procedure.

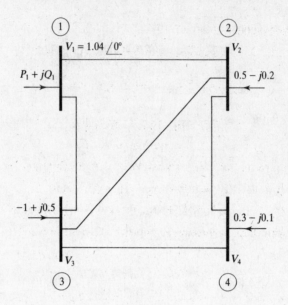

Fig. 8-9.

From (8.21),

$$V_2^{(2)} = \frac{1}{Y_{22}} \left[\frac{P_2 - jQ_2}{(V_2^{(1)})^*} - Y_{21}V_1 - Y_{23}V_3^{(1)} - Y_{24}V_4^{(1)} \right]$$

Substituting the given numerical values and $V_3^{(1)}$ and $V_4^{(1)}$ (which may be determined in the same way as $V_2^{(1)}$ in Problem 8.10), we obtain

$$V_2^{(2)} = \frac{1}{Y_{22}} \left[\frac{0.5 + j0.2}{1.019 + j0.046} - 1.04(-2 + j6) - (-0.666 + j2.0)(1.028 - j0.087) \right.$$

$$\left. - (-1 + j3)(1.025 - j0.0093) \right]$$

$$= \frac{4.0862 - j11.6119}{3.666 - j11.0} = 1.061 + j0.0179$$

$$= 1.0616\underline{/0.97°} \text{ pu}$$

8.12   Solve the following equations by the Newton–Raphson method:

$$x_1^2 - 4x_2 - 4 = 0$$
$$2x_1 - x_2 - 2 = 0$$

Let $x_1^{(0)} = 1$ and $x_2^{(0)} = -1$ be the starting point for the first iteration. Then

$$f_1(x_1^{(0)}, x_2^{(0)}) = 1 + 4 - 4 = 1$$
$$f_2(x_1^{(0)}, x_2^{(0)}) = 2 + 1 - 2 = 1$$

and the partial derivatives evaluated at $x_1^{(0)}$ and $x_2^{(0)}$ are

$$\frac{\partial f_1}{\partial x_1} = 2x_1 = 2 \qquad \frac{\partial f_2}{\partial x_1} = 2$$

$$\frac{\partial f_1}{\partial x_2} = -4 \qquad \frac{\partial f_2}{\partial x_2} = -1$$

Now (8.28) yields the equations

$$f_1(x_1^{(0)}, x_2^{(0)}) + \Delta x_1^{(0)} \frac{\partial f_1}{\partial x_1} + \Delta x_2^{(0)} \frac{\partial f_1}{\partial x_2} = 0$$

$$f_2(x_1^{(0)}, x_2^{(0)}) + \Delta x_1^{(0)} \frac{\partial f_2}{\partial x_1} + \Delta x_2^{(0)} \frac{\partial f_2}{\partial x_2} = 0$$

and substitution yields

$$1 + 2\Delta x_1 - 4\Delta x_2 = 0$$
$$1 + 2\Delta x_1 - \Delta x_2 = 0$$

Simultaneous solution gives us $\Delta x_1^{(0)} = -0.5$ and $\Delta x_2^{(0)} = 0$. Thus, better estimates of x_1 and x_2 are

$$x_1^{(1)} = x_1^{(0)} + \Delta x_1 = 1 - 0.5 = 0.5$$

and

$$x_2^{(1)} = x_2^{(0)} + \Delta x_2 = -1 + 0 = -1.0$$

Proceeding as above with these new estimates, we find that a second and third iteration yield

$$x_1^{(2)} = 0.5357 \quad \text{and} \quad x_2^{(2)} = -0.9286$$
$$x_1^{(3)} = 0.5359 \quad \text{and} \quad x_2^{(3)} = -0.9282$$

Clearly, such problems are solved most conveniently with a digital computer.

8.13 For the system shown in Fig. 8-10,

$$\mathbf{Y}_{\text{bus}} = \begin{bmatrix} 24.23\underline{/-75.95°} & 12.13\underline{/104.04°} & 12.13\underline{/104.04°} \\ 12.13\underline{/104.04°} & 24.23\underline{/-75.95°} & 12.13\underline{/104.04°} \\ 12.13\underline{/104.04°} & 12.13\underline{/104.04°} & 24.23\underline{/-75.95°} \end{bmatrix} \text{pu}$$

Given the per-unit voltages and power as shown, determine V_2 by the Newton–Raphson method.

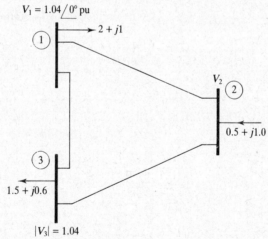

$V_1 = 1.04\underline{/0°}$ pu

V_2

$2 + j1$

$0.5 + j1.0$

$1.5 + j0.6$

$|V_3| = 1.04$

Fig. 8-10.

Let $V_2^{(0)} = 1\underline{/0°}$ pu and $\delta_3^{(0)} = 0$. Then from (8.33),

$$P_2^{(0)} = |V_2^{(0)}| \, |V_1^{(0)}| \, |Y_{21}| \cos(\theta_{21} + \delta_1^{(0)} - \delta_2^{(0)}) + |V_2^{(0)}|^2 \, |Y_{22}| \cos\theta_{22}$$
$$+ |V_2^{(0)}| \, |V_3^{(0)}| \, |Y_{23}| \cos(\theta_{23} + \delta_3^{(0)} - \delta_2^{(0)})$$
$$= (1)(1.04)(12.31) \cos 104.04° + (1)(24.23) \cos(-75.95°)$$
$$+ (1)(1.04)(12.31) \cos 104.04°$$
$$= -0.33 \text{ pu}$$

Similarly,

$$
\begin{aligned}
P_3^{(0)} &= |V_3^{(0)}|\,|V_1^{(0)}|\,|Y_{31}|\cos(\theta_{31} + \delta_1^{(0)} - \delta_3^{(0)}) + |V_3^{(0)}|\,|V_2^{(0)}|\,|Y_{32}| \\
&\quad \times \cos(\theta_{32} + \delta_2^{(0)} - \delta_3^{(0)}) + |V_3^{(0)}|^2\,|Y_{33}|\cos\theta_{33} \\
&= (1.04)(1.04)(12.31)\cos 104.04° + (1.04)(12.31)\cos 104.04° \\
&\quad + (1.04)^2(24.23)\cos(-75.95°) \\
&= 0.026\ \text{pu}
\end{aligned}
$$

Also, from (8.34),

$$
\begin{aligned}
Q_2^{(0)} &= -|V_2^{(0)}|\,|V_1^{(0)}|\,|Y_{21}|\sin(\theta_{21} + \delta_1^{(0)} - \delta_2^{(0)}) - |V_2^{(0)}|^2\,|Y_{22}|\sin\theta_{22} \\
&\quad - |V_2^{(0)}|\,|V_3^{(0)}|\sin(\theta_{23} + \delta_2^{(0)} - \delta_3^{(0)}) \\
&= -(1)(1.04)(12.31)\sin 104.04° - (1)(24.23)\sin(-75.95°) \\
&\quad - (1)(1.04)(12.31)\sin 104.04° \\
&= -1.33\ \text{pu}
\end{aligned}
$$

Now, from (8.35)

$$
\Delta P_2^{(0)} = 0.5 - (-0.33) = 0.83\ \text{pu}
$$
$$
\Delta P_3^{(0)} = -1.5 - 0.026 = -1.526\ \text{pu}
$$

Similarly, from (8.36),

$$
\Delta Q_2^{(0)} = 1 - (-1.33) = 2.33\ \text{pu}
$$

For the given three-bus system (with V_3 known), (8.37) becomes

$$
\begin{bmatrix} \Delta P_2^{(0)} \\ \Delta P_3^{(0)} \\ \Delta Q_2^{(0)} \end{bmatrix} =
\begin{bmatrix}
\dfrac{\partial P_2}{\partial \delta_2} & \dfrac{\partial P_2}{\partial \delta_3} & \dfrac{\partial P_2}{\partial |V_2|} \\[2mm]
\dfrac{\partial P_3}{\partial \delta_2} & \dfrac{\partial P_3}{\partial \delta_3} & \dfrac{\partial P_3}{\partial |V_2|} \\[2mm]
\dfrac{\partial Q_2}{\partial \delta_2} & \dfrac{\partial Q_2}{\partial \delta_3} & \dfrac{\partial Q_2}{\partial |V_2|}
\end{bmatrix}
\begin{bmatrix} \Delta \delta_2^{(0)} \\ \Delta \delta_3^{(0)} \\ \Delta |V_2^{(0)}| \end{bmatrix}
\tag{1}
$$

Differentiating (8.33) and (8.34) and substituting the numerical values yields, from (1) above,

$$
\begin{bmatrix} 0.83 \\ -1.526 \\ 2.33 \end{bmatrix} =
\begin{bmatrix}
24.47 & -12.23 & 5.64 \\
-12.23 & 24.95 & -3.05 \\
-6.11 & 3.05 & 22.54
\end{bmatrix}
\begin{bmatrix} \Delta \delta_2^{(0)} \\ \Delta \delta_3^{(0)} \\ \Delta |V_2^{(0)}| \end{bmatrix}
\tag{2}
$$

Or

$$
\begin{aligned}
\begin{bmatrix} \Delta \delta_2^{(0)} \\ \Delta \delta_3^{(0)} \\ \Delta |V_2^{(0)}| \end{bmatrix} &=
\begin{bmatrix}
24.47 & -12.23 & 5.64 \\
-12.23 & 24.95 & -3.05 \\
-6.11 & 3.05 & 22.54
\end{bmatrix}^{-1}
\begin{bmatrix} 0.83 \\ -1.526 \\ 2.33 \end{bmatrix} \\[3mm]
&= \begin{bmatrix}
0.05179 & 0.02653 & -0.00937 \\
0.02666 & 0.05309 & 0.00051 \\
0.01043 & -0.00001 & 0.04176
\end{bmatrix}
\begin{bmatrix} 0.83 \\ -1.526 \\ 2.33 \end{bmatrix}
\end{aligned}
$$

Solving for $\Delta |V_2^{(0)}|$ in (2) gives

$$
\Delta |V_2^{(0)}| = (0.01043)(0.83) + (0.00001)(1.526) + (0.04176)(2.33) = 0.106\ \text{pu}
$$

Thus, $|V_2^{(1)}| = 1 + 0.106 = 1.106\ \text{pu}$

This procedure is repeated until, upon convergence, we obtain $V_2 = 1.081\underline{/-1.37°}$ pu.

Supplementary Problems

8.14 For a system of the type shown in Fig. 8-7, $|V_1| = 1.0$ pu and $|V_2| = 1.1$ pu. The complex power outputs at the two buses are equal; that is, $S_1 = S_2 = 3 + j4$ pu, and the real power supplied by each generator is 5.0 pu. If the line reactance is 0.08 pu, calculate the load on each bus.

Ans. $8 + j5.71$ pu

8.15 In the system of Fig. 8-2, $V_1 = 1\underline{/0^\circ}$ pu, $Z_\ell = (0.10 + j0.10)$ pu, and $P_2 + jQ_2 = (0.1 + j0.1)$ pu. Calculate V_2.

Ans. $0.9796\underline{/0^\circ}$ pu

8.16 Determine the real and reactive power on bus 1 in Problem 8.15.

Ans. $P_1 = Q_1 = 0.10208$ pu

8.17 How much reactive power must be supplied at bus 2 in Problem 8.15 so that $|V_2| = 1.1$ pu?

Ans. 10.93 pu

8.18 For the system shown in Fig. 8-2, $Z_\ell = (0.2 + j0.6)$ pu, $V_1 = 1.1\underline{/0^\circ}$ pu, and $P_2 + jQ_2 = (1 + j0.4)$ pu. Calculate V_2.

Ans. $1.371\underline{/-19.46^\circ}$ pu

8.19 Determine the complex power on bus 1 in Problem 8.18.

Ans. $(0.65 + j0.57)$ pu

8.20 A generator is connected to a system as shown by the equivalent circuit of Fig. 8-11. If $V_t = 0.97\underline{/0^\circ}$ pu, calculate the complex power delivered by the generator. Also determine E_g.

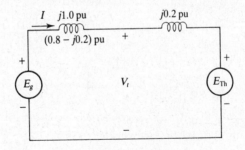

Fig. 8-11.

Ans. $(0.776 + j0.194)$ pu; $1.42\underline{/34.3^\circ}$ pu

8.21 Repeat Problem 8.20 if $|V_t| = 1.0$ pu and the real power delivered by the generator remains unchanged. Assume E_{Th} to remain constant.

Ans. $(0.776 + j0.346)$ pu; $1.55\underline{/29.7^\circ}$ pu

8.22 Solve the following system of equations by the Gauss–Seidel method, starting with $x(0) = y(0) = 0$:

$$10x + 5y = 6$$
$$2x + 9y = 3$$

Ans. $x = 0.4875$; $y = 0.2250$ after six iterations

8.23 Evaluate x in the equation $x + \sin x = 2$ by the Gauss–Seidel method. Start with $x(0) = 0$.

    Ans. 1.106

8.24 Rework Problem 8.23 using the Newton–Raphson method, and compare the number of iterations required to achieve convergence by the two methods.

    Ans. Three iterations versus ten iterations.

8.25 Determine the values for V_3 and V_4 in the system of Fig. 8-9 and Problem 8.10, as produced by the first iteration of the Gauss–Seidel procedure.

    Ans. $V_3^{(1)} = 1.028 - j0.087$ pu; $V_4^{(1)} = 1.025 - j0.0093$ pu

8.26 Given the following set of equations,

$$0.6270I_1 + 0.1930I_2 + 0.0100I_3 = 1.0$$
$$0.1930I_1 + 0.4840I_2 + 0.1711I_3 = 1.0$$
$$0.0100I_1 + 0.1711I_2 + 0.6960I_3 = 1.0$$

start with $I_1^{(0)} = I_2^{(0)} = I_3^{(0)} = 1$ and find $I_1^{(1)}$, $I_1^{(6)}$, and $I_2^{(3)}$ by the Gauss–Seidel method.

    Ans. $I_1^{(1)} = 1.2711$; $I_2^{(3)} = 1.1847$; $I_1^{(6)} = 1.2125$

8.27 In a five-bus system $Y_{21} = Y_{23} = 0$, $Y_{22} = 7.146\underline{/-84.6°}$ pu, $Y_{24} = 2.490\underline{/95.1°}$ pu, and $Y_{25} = 4.980\underline{/95.1°}$ pu. Determine $V_2^{(2)}$ by the Gauss–Seidel method if $P_2 - jQ_2 = -2 + j0.7$. Begin with $V_1^{(0)} = V_2^{(0)} = V_3^{(0)} = V_4^{(0)} = V_5^{(0)} = 1\underline{/0°}$ pu.

    Ans. $0.875\underline{/-15.7°}$ pu

8.28 For the system shown in Fig. 8-12, with bus 3 as reference bus, the bus impedance matrix is

$$\mathbf{Z}_{bus} = \begin{bmatrix} 1.33 + j1.33 & 1 + j1 \\ 1 + j1 & 1.5 + j1.5 \end{bmatrix} \times 10^{-2} \text{ pu}$$

Start with $V_1^{(0)} = V_2^{(0)} = 1.05\underline{/0°}$, and solve for V_1 and V_2 by the Gauss–Seidel method.

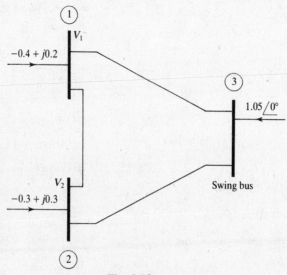

Fig. 8-12.

    Ans. $1.04736\underline{/-7.29°}$ pu; $1.0480\underline{/-7.81°}$ pu

8.29   Rework Problem 8.28 using the Newton–Raphson method.

8.30   For a three-bus system similar to that of Fig. 8-12,

$$\mathbf{Y}_{bus} = \begin{bmatrix} 1.6 - j8 & -0.8 + j4 & -0.8 + j4 \\ -0.8 + j4 & 1.6 - j8 & -0.8 + j4 \\ -0.8 + j4 & -0.8 + j4 & 1.6 - j8 \end{bmatrix} \text{pu}$$

Also, $V_1 = 1\underline{/0°}$ and $P_2 - jQ_2 = -0.8 + j0.6$. Determine $V_2^{(2)}$ in the Gauss–Seidel procedure.

Ans. $0.95\underline{/-4.59°}$

Chapter 9

Power System Operation and Control

Of the numerous aspects of power system operation and control, we shall consider only the economical operation of power systems and the control of load frequency, generator voltage, and the turbine governor.*

9.1 ECONOMIC DISTRIBUTION OF LOAD BETWEEN GENERATORS

Within a power plant, a number of ac generators generally operate in parallel. For the economic operation of the plant, the total load must be appropriately shared by the generating units. Because fuel cost is the major factor in determining economic operation, curves like that of Fig. 9-1 are important to power-plant operation. Note in the figure that the inverse slope of the curve at any point is the fuel efficiency of the generating unit operating at that point. Maximum fuel efficiency occurs at the point at which the line from the origin is tangent to the curve. Point A in Fig. 9-1 is such a point for a unit having the input–output characteristic of Fig. 9-1; there, an output of 250 MW requires an input of approximately 2.1×10^9 Btu/h. Or, we may say that the *fuel requirement* is 8.4×10^6 Btu/MWh.

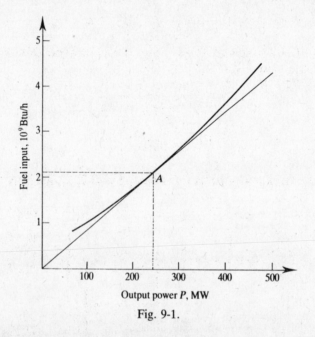

Fig. 9-1.

To obtain the most economical load distribution between two units, we must determine the incremental cost corresponding to a partial shift of load between the units. We first convert the fuel requirement into a dollar cost per megawatthour. Then the incremental cost is determined from the slopes of the input–output curves (Fig. 9-1) for the two units. From Fig. 9-1, for each unit,

$$\text{Incremental fuel cost} = \frac{dF}{dP} \quad \text{(in dollars per megawatthour)} \quad (9.1)$$

---

* Discussions in this Chapter follow W. D. Stevenson, Jr., *Elements of Power System Analysis*, 4th ed., McGraw-Hill, 1982.

where F = input in dollars per hour, and P = output in megawatts. At a given output, this incremental fuel cost is the additional cost of increasing the output by 1 MW. (See Problems 9.2 and 9.3.)

In a plant having two operating units, generally the incremental fuel cost of one unit will be higher than that of the other. For the most economic operation, load should be transferred from the unit with the higher incremental cost to the unit with the lower incremental cost, until the incremental costs of the two become equal. In a plant with several units, the criterion for load division is that all units must operate at the same incremental fuel cost. (This conclusion may be derived mathematically, as is shown in Problem 9.5.) If a plot of dF_k/dP_k versus P_T for each unit is linear, then λ may be plotted versus P_T to determine the optimum value of λ, where F_k is the input to unit k in dollars per hour, and $\lambda = dF_k/dP_k$ is the incremental fuel cost for unit k in dollars per megawatthour. (λ is also known as the *Lagrange multiplier*; see Problem 9.6.)

9.2 EFFECT OF TRANSMISSION-LINE LOSS

To include the effect of transmission-line losses on economical system operation, we must express these losses as a function of plant power output. Figure 9-2 shows two plants connected to a three-phase load. The total transmission loss (for all three phases) is given by

$$P_{\text{loss}} = 3(|I_1|^2 R_1 + |I_2|^2 R_2 + |I_3|^2 R_3) \tag{9.2}$$

where the R's are the per-phase resistances of the lines, and where

$$|I_3| = |I_1 + I_2| = |I_1| + |I_2| \tag{9.3}$$

if we assume that I_1 and I_2 are in phase. These currents may be expressed in terms of P_1 and P_2, the respective plant outputs, as

$$|I_1| = \frac{P_1}{\sqrt{3}\,|V_1|\cos\phi_1} \tag{9.4}$$

and

$$|I_2| = \frac{P_2}{\sqrt{3}\,|V_2|\cos\phi_2} \tag{9.5}$$

where $\cos\phi_1$ and $\cos\phi_2$ are the power factors at buses 1 and 2, respectively.

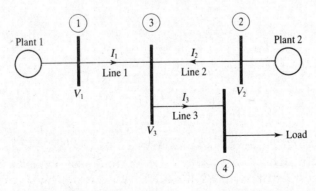

Fig. 9-2.

Equations (9.2) through (9.5) may be combined to yield

$$P_{\text{loss}} = P_1^2 B_{11} + 2P_1 P_2 B_{12} + P_2^2 B_{22} \tag{9.6}$$

where B_{11}, B_{12}, and B_{22} are called *loss coefficients* or *B coefficients* and are given by

$$B_{11} = \frac{R_1 + R_3}{|V_1|^2 \cos^2 \phi_1} \tag{9.7}$$

$$B_{12} = \frac{R_3}{|V_1|\,|V_2| \cos \phi_1 \cos \phi_2} \tag{9.8}$$

and

$$B_{22} = \frac{R_2 + R_3}{|V_2|^2 \cos^2 \phi_2} \tag{9.9}$$

For a system of n plants, (9.6) may be generalized to

$$P_{\text{loss}} = \sum_{k=1}^{n} \sum_{m=1}^{n} P_k P_m B_{km} \tag{9.10}$$

where $B_{mk} = B_{km}$.

9.3 LOAD DISTRIBUTION BETWEEN PLANTS

In this section we combine the method of Section 9.1 with the results of Section 9.2 to obtain an economical allocation of load among a number of power plants. For a system of n plants, the total cost of fuel per hour is

$$F_{\text{total}} = \sum_{k=1}^{n} F_k \qquad \text{dollars/h} \tag{9.11}$$

and the total power output is

$$P_{\text{total}} = \sum_{k=1}^{n} P_k \qquad \text{MW} \tag{9.12}$$

With transmission losses, we must have

$$P_{\text{total}} = P_R + P_{\text{loss}} \tag{9.13}$$

where P_R and P_{loss} are, respectively, the total power received by the load and lost in transmission. For a given (constant), P_R, $dP_R = 0$. Thus, (9.12) and (9.13) yield

$$\sum_{k=1}^{n} dP_k - dP_{\text{loss}} = 0 \tag{9.14}$$

In addition, when the load is allocated among the n plants for minimum fuel cost, $dF_{\text{total}} = 0$. Then

$$dF_{\text{total}} = \sum_{k=1}^{n} \frac{\partial F_{\text{total}}}{\partial P_k} dP_k = 0 \tag{9.15}$$

Also

$$dP_{\text{loss}} = \sum_{k=1}^{n} \frac{\partial P_{\text{loss}}}{\partial P_k} dP_k \tag{9.16}$$

Substituting dP_{loss} from (9.16) into (9.14), multiplying the result by λ, and subtracting that result from the right-hand equality in (9.15) yield

$$\sum_{k=1}^{n} \left[\left(\frac{\partial F_{\text{total}}}{\partial P_k} + \lambda \frac{\partial P_{\text{loss}}}{\partial P_k} - \lambda \right) dP_k \right] = 0 \tag{9.17}$$

This equation holds if

$$\frac{\partial F_{\text{total}}}{\partial P_k} + \lambda \frac{\partial P_{\text{loss}}}{\partial P_k} - \lambda = 0 \qquad \text{for all } k = 1, 2, \ldots, n \tag{9.18}$$

Now, because

$$\frac{\partial F_{\text{total}}}{\partial P_k} = \frac{dF_k}{dP_k} \tag{9.19}$$

condition (9.18) may be written as

$$\frac{dF_k}{dP_k} L_k = \lambda \qquad \text{for } k = 1, 2, \ldots, n \tag{9.20}$$

where L_k, called the *penalty factor* of the kth plant, is given by

$$L_k = \frac{1}{1 - \partial P_{\text{loss}}/\partial P_k} \tag{9.21}$$

Condition (9.20) implies that the system fuel cost is minimized when the incremental fuel cost for each plant, multiplied by its penalty factor, is the same throughout the system, that is, when

$$\frac{dF_1}{dP_1} L_1 = \frac{dF_2}{dP_2} L_2 = \cdots = \frac{dF_n}{dP_n} L_n = \lambda \tag{9.22}$$

To determine the L_k we have, from (9.10),

$$\frac{\partial P_{\text{loss}}}{\partial P_k} = \frac{\partial}{\partial P_k} \left(\sum_{k=1}^{n} \sum_{m=1}^{n} P_k P_m B_{km} \right) = 2 \sum_{m=1}^{n} P_m B_{mk} \tag{9.23}$$

The simultaneous equations represented by (9.20) can be solved if a value is assumed for λ.

9.4 POWER SYSTEM CONTROL

A number of automatic controls are used in present-day power systems. These include devices that control the generator voltage, the turbine governor, and the load frequency; there are also computer controls to ensure economic power flow and to control reactive power, among other power-system variables.

Generator voltage control is accomplished by controlling the exciter voltage. The block-diagram representation of a closed-loop automatic voltage regulating system is shown in simplified form in Fig. 9-3. (Numerous other forms of generator voltage control also exist.) In Fig. 9-3, the open-loop transfer function $G(s)$ is given by

$$G(s) = \frac{k}{(1 + T_a s)(1 + T_e s)(1 + T_f s)} \tag{9.24}$$

where T_a, T_e, and T_f are, respectively, the time constants associated with the amplifier, the exciter, and the generator field, and the open-loop gain k is

$$k = k_a k_e k_f \tag{9.25}$$

Sudden changes in the load cause the turbine speed and, consequently, the generator frequency to change as well. The change in turbine speed occurs when the generator electromagnetic torque no longer equals the turbine (or other prime mover) mechanical torque. Thus, the change Δf in the generator frequency may be used as a control signal for controlling the turbine mechanical output power. The change in the turbine output power as a function of a change in the generator frequency is given by

$$\Delta P_m = \Delta P_{\text{ref}} - \frac{1}{R} \Delta f \tag{9.26}$$

where ΔP_m and ΔP_{ref} are, respectively, the changes in the turbine output power and the reference power (as determined by the governor setting), and R is known as the *regulation constant*.

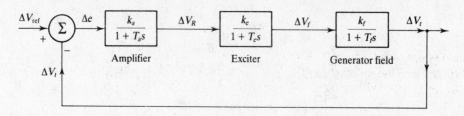

Fig. 9-3.

Figure 9-4 shows the block diagram for a turbine-governor control system; we assume the system to be linear, and the governor and turbine-generator to be first-order devices.

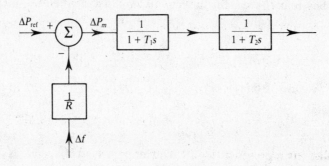

Fig. 9-4.

In the preceding, we have implied that the accelerations and decelerations of the generator rotor are controlled by the turbine governor. However, the frequency deviation Δf still remains if $\Delta P_{\text{ref}} = 0$. This frequency deviation can be reduced to zero by a process called *load-frequency control* (or LFC). The LFC process then also controls the power flow on the tie line. Thus, via LFC, each interconnected area of a power system maintains the power flow out of that area at its scheduled value, in effect absorbing its own load variations.

To establish the pertinent control strategy for LFC, we define the *area control error* (ACE) as

$$\text{ACE} = \Delta P_{\text{tie}} + B_f \, \Delta f \qquad (9.27)$$

where ΔP_{tie} is the deviation of the tie-line power flow out of the area from the scheduled power flow, Δf is the frequency error, and B_f is known as the *frequency bias constant*.

The change ΔP_{ref} in the reference power setting of the load-frequency-controlled turbine governor is proportional to the integral of ACE. Thus,

$$\Delta P_{\text{ref}} = -K \int \text{ACE} \, dt \qquad (9.28)$$

where K is a constant. The minus sign in (9.28) implies that if either the net power flow out of the area or the frequency is low, then ACE is negative and the area should increase its generation.

If an area contains n generating units, we may write (9.26) as

$$\Delta P_{m(\text{total})} = \sum_{k=1}^{n} \Delta P_{mk} = \sum_{k=1}^{n} \Delta P_{\text{ref}k} - \left(\sum_{k=1}^{n} \frac{1}{R_k} \right) \Delta f = \Delta P_{\text{ref(total)}} - \beta \, \Delta f \qquad (9.29)$$

where β is known as the *area frequency-response characteristic* and is given by

$$\beta = \sum_{k=1}^{n} \frac{1}{R_k} \qquad (9.30)$$

Also,

$$\Delta P_{\text{ref(total)}} = \sum_{k=1}^{n} \Delta P_{\text{ref}k} \qquad (9.31)$$

and Δf remains the same for each unit.

To summarize, (9.27) through (9.29) govern the LFC of the system. Problems 9.19 and 9.20 illustrate the procedure with numerical examples.

Solved Problems

9.1 Use Fig. 9-1 to find the fuel requirements for outputs of (a) 100 MW and (b) 400 MW. Thus verify that point A is probably the maximum fuel-efficiency point.

(a) From Fig. 9-1, at 100 MW output, the fuel input is approximately 1×10^9 Btu/h. Hence,

$$\text{Fuel requirement} = \frac{1 \times 10^9}{100} = 10 \times 10^6 \text{ Btu/MWh}$$

(b) Similarly, at 400 MW output, the fuel input is approximately 3.6×10^9 Btu/h. Then

$$\text{Fuel requirement} = \frac{3.6 \times 10^9}{400} = 9.0 \times 10^6 \text{ Btu/MWh}$$

Clearly, both values are greater than that for point A.

9.2 A certain amount of coal costs \$1.20 and produces 10^6 Btu of energy as fuel for a generating unit. If the input–output characteristic of the unit is that shown in Fig. 9-1, determine the incremental fuel cost at point A.

$$\text{Slope at } A = \frac{(2.2 - 2.0)10^9}{(260 - 234)10^6} = 7.7 \text{ Btu/MWh}$$

Thus,   Incremental cost = $7.7 \times 1.20 = \$9.24/\text{MWh}$

9.3 Convert the curve of Fig. 9-1 to a plot of incremental fuel cost versus output power, given a fuel cost of \$1.50 per 10^6 Btu.

We plot the incremental fuel cost by finding the slope of the input–output curve (Fig. 9-1) for several values of output power and plotting slope × cost per Btu against output power. (See Problem 9.2.) Hence we obtain the curve shown in Fig. 9-5.

9.4 Approximate the curve obtained in Problem 9.3 with a straight line, and obtain an equation for the straight line. Use it to determine the incremental cost at 250 MW.

The approximation is shown in Fig. 9-5. The line has an intercept of \$6.25/MWh and a slope of 0.0226. Thus, the required equation is

$$\frac{dF}{dP} = 0.0226P + 6.25 \qquad (1)$$

Substituting $P = 250$ in (1) yields

$$\text{Incremental cost} = \frac{dF}{dP} = 0.0226 \times 250 + 6.25$$

$$= \$11.9/\text{MWh}$$

9.5 Show that, for the most economic operation of a power plant having several generating units,

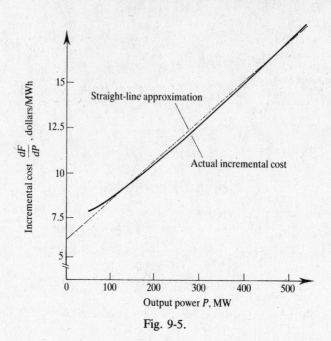

Fig. 9-5.

the load must be divided among the units such that they all operate at the same incremental cost λ.

If there are n units in the plant, the total input fuel cost F_{total}, in dollars per hour, is given by

$$F_{\text{total}} = \sum_{k=1}^{n} F_k \tag{1}$$

The total output power P_{total}, in megawatts, may be written as

$$P_{\text{total}} = \sum_{k=1}^{n} P_k \tag{2}$$

For a given P_{total}, F_{total} is a minimum when $dF_{\text{total}} = 0$, that is, when

$$dF_{\text{total}} = \sum_{k=1}^{n} \frac{\partial F_{\text{total}}}{\partial P_k} dP_k = 0 \tag{3}$$

Since P_{total} is constant, $dP_{\text{total}} = 0$. Then (2) yields

$$dP_{\text{total}} = \sum_{k=1}^{n} dP_k = 0 \tag{4}$$

Multiplying (4) by λ and subtracting the result from (3) yield

$$\sum_{k=1}^{n} \left[\left(\frac{\partial F_{\text{total}}}{\partial P_k} - \lambda \right) dP_k \right] = 0 \tag{5}$$

The sum in (5) will be zero if each term in parentheses is zero. Moreover, for each unit, $\partial F_{\text{total}}/\partial P_k = dF_k/dP_k$, because a change in a unit's power output affects only that unit's fuel cost. Hence

$$\frac{\partial F_{\text{total}}}{\partial P_k} = \frac{dF_k}{dP_k} = \lambda$$

and the required condition is

$$\frac{dF_1}{dP_1} = \frac{dF_2}{dP_2} = \cdots = \frac{dF_n}{dP_n} = \lambda$$

9.6　Graphs of the incremental fuel costs (in dollars per megawatthour) for two generating units in a power plant are shown in Fig. 9-6. These graphs are linear. The plant output ranges from 240 MW to 1000 MW over a 24-h period. During this period the load on each unit varies from 120 MW to 600 MW. Plot a curve of incremental fuel cost λ versus plant output for minimum-fuel-cost operation.

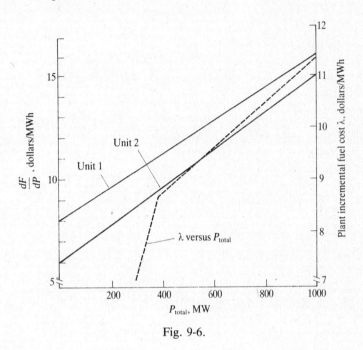

Fig. 9-6.

From Fig. 9-6, we obtain

$$\frac{dF_1}{dP_1} = 0.008P_1 + 8 \tag{1}$$

and

$$\frac{dF_2}{dP_2} = 0.009P_2 + 6 \tag{2}$$

At 120 MW, $dF_1/dP_1 = 8.96$ and $dF_2/dP_2 = 7.08$. Therefore, until dF_2/dP_2 has risen to 8.96, unit 2 should take all the additional load above 120 MW. Using (2), we find that dF_2/dP_2 is equal to 8.96 when $P_2 = 328.9$ MW, at which value

$$P_{total} = P_1 + P_2 = 120 + 328.9 = 448.9 \text{ MW}$$

These values give us the second row of Table 9-1. Similar computations give the remaining rows of the table, whose values are plotted in Fig. 9-6 (dashed lines).

TABLE 9-1

λ, dollars/MWh	P_1, MW	P_2, MW	P_{total}, MW
7.17	120	130	250
8.96	120	328.9	448.9
9.6	200	400	600
10	250	444.4	694.4
10.4	300	488.9	788.9
11.2	400	577.77	977.7

9.7 For maximum demand from the plant of Problem 9.6, determine how the load should be shared by the two generating units for minimum fuel cost.

The maximum load is

$$P_{\text{total}} = 1000 = P_1 + P_2 \tag{1}$$

For minimum fuel cost, (1) and (2) of Problem 9.6 give

$$0.008P_1 + 8 = 0.009P_2 + 6 \tag{2}$$

Solving (1) and (2) for P_1 and P_2 yields

$$P_1 = 411.76 \text{ MW} \quad \text{and} \quad P_2 = 588.24 \text{ MW}$$

9.8 Determine the incremental fuel cost for each unit for the conditions of Problem 9.7.

The incremental fuel cost λ is the same for both units. With $P_1 = 411.76$ MW, (1) of Problem 9.6 gives

$$\lambda = 0.008 \times 411.76 + 8 = \$11.29/\text{MWh}$$

which agrees with the last row of Table 9-1 (Problem 9.6).

9.9 For the maximum power output of the plant of Problem 9.6, calculate the saving per hour in fuel cost under economic (optimal) operation, as compared to operation with the load equally divided between the two units.

For economic operation we have, from Problem 9.7, $P_1 = 411.76$ MW and $P_2 = 588.24$ MW. If $P_1 = P_2 = 500$ MW, the increase in cost per hour for operating unit 1 is

$$\Delta_1 = \int_{411.76}^{500} (0.008P_1 + 8) \, dP_1 = \$1027.73/\text{h}$$

For unit 2 we have a decrease in cost:

$$\Delta_2 = \int_{588.24}^{500} (0.009P_2 + 6) \, dP = -\$961.56/\text{h}$$

The net increase in operating cost is therefore $1027.73 - 961.56 = \$66.17/\text{h}$

9.10 Find the loss coefficients for the system shown in Fig. 9-2 from the following data, in which all numerical quantities are per-unit values: $I_1 = 0.8\underline{/0°}$, $I_2 = 0.9\underline{/0°}$, $V_3 = 1.1\underline{/0°}$, $Z_1 = Z_2 = 0.06 + j0.20$, and $Z_3 = 0.04 + j0.06$.

From Fig. 9-2 and the given data,

$$V_1 = 1.1 + (0.8\underline{/0°})(0.06 + j0.20) = 1.148 + j0.16$$
$$V_2 = 1.1 + (0.9\underline{/0°})(0.06 + j0.20) = 1.156 + j0.18$$

Hence, $|V_1| \cos \phi_1 = 1.148$ and $|V_2| \cos \phi_2 = 1.156$. Now (9.7) through (9.9) yield

$$B_{11} = \frac{0.06 + 0.04}{(1.148)^2} = 0.0759 \text{ pu}$$

$$B_{12} = \frac{0.04}{(1.148)(1.156)} = 0.0301 \text{ pu}$$

$$B_{22} = \frac{0.06 + 0.04}{(1.156)^2} = 0.0748 \text{ pu}$$

9.11 Calculate the transmission loss for the system of Problem 9.10.

We have
$$P_1 = \text{Re}\,[(0.8\underline{/0^\circ})(1.148 + j0.16)] = 0.9184 \text{ pu}$$
$$P_2 = \text{Re}\,[(0.9\underline{/0^\circ})(1.156 + j0.18)] = 1.0404 \text{ pu}$$

Substituting these values and the B coefficients of Problem 9.10 in (9.6) gives

$$P_{\text{loss}} = (0.8384)^2(0.0910) + 2(0.8384)(0.904)(0.0360) + (0.9504)^2(0.0897) = 0.2024 \text{ pu}$$

9.12 In a two-plant system, the entire load is located at plant 2, which is connected to plant 1 by a transmission line. Plant 1 supplies 100 MW of power with a corresponding transmission loss of 5 MW. Calculate the penalty factors for the two plants.

Since all the load is at plant 2, varying P_2 does not affect the transmission loss P_{loss}. Thus, from (9.6),
$$P_{\text{loss}} = 5 = P_1^2 B_{11} = 10^4 B_{11}$$

so that $B_{11} = 5 \times 10^{-4}\,\text{MW}^{-1}$. Moreover, this expression for P_{loss} yields

$$\frac{\partial P_{\text{loss}}}{\partial P_1} = 2P_1 B_{11} = 2(100)(5 \times 10^{-4}) = 0.1$$

Then, from (9.21),
$$L_1 = \frac{1}{1 - 0.1} = 1.111$$

Similarly, because $\partial P_{\text{loss}}/\partial P_2 = 0$, we have $L_2 = 1$.

9.13 For the system of Problem 9.12, $\lambda = \$15/\text{MWh}$, and the incremental fuel costs for the two plants are given by

$$\frac{dF_1}{dP_1} = 0.01P_1 + 10 \quad \text{and} \quad \frac{dF_2}{dP_2} = 0.02P_2 + 12$$

in dollars per megawatthour. How much power should be generated at each plant for minimal total fuel cost?

From (9.22) and Problem 9.12,
$$\frac{dF_1}{dP_1} = \lambda$$

or
$$(0.01P_1 + 10)1.111 = 15$$

from which $P_1 = 350$ MW. Similarly,

$$\frac{dF_2}{dP_2} L_2 = (0.02P_2 + 12)1 = 15$$

from which $P_2 = 150$ MW.

9.14 For the operating conditions obtained in Problem 9.13, determine the dollar savings that would be realized by coordinating the transmission loss rather than neglecting its effect.

With the transmission not coordinated, we would have (based on Section 9.1) $dF_1/dP_1 = dF_2/dP_2$ for economic operation. Hence we would have

$$0.01P_1 + 10 = 0.02P_2 + 12 \tag{1}$$

From Problem 9.13, we know that the load requires

$$P_1 + P_2 - P_{\text{loss}} = 350 + 150 - P_1^2 B_{11}$$
$$= 500 - 61.25 = 438.75 \text{ MW}$$

Then with the transmission loss *not* coordinated, we would have

$$P_1 + P_2 - 5 \times 10^{-4}P_1^2 = 438.75 \tag{2}$$

Solving (1) and (2) simultaneously yields $P_1 = 417$ MW and $P_2 = 108.5$ MW.

Comparing these results with the results of Problem 9.13, we see that the load on plant 1 is increased from 350 MW to 417 MW; hence its fuel cost increases by

$$\int_{350}^{417} (0.01P_1 + 10)\, dP_1 = \$926.945/\text{h}$$

The load on plant 2 is decreased from 150 MW to 108.5 MW; hence its fuel cost decreases by

$$-\int_{150}^{108.5} (0.02P_2 + 12)\, dP_2 = \$605.277/\text{h}$$

The saving with loss coordination is thus $926.945 - 605.277 = \$321.67/\text{h}$.

9.15 For the system shown in Fig. 9-3, what is the minimum open-loop gain such that the steady-state error Δe_{ss} does not exceed 1 percent?

From Fig. 9-3,

$$\frac{\Delta e}{\Delta V_{\text{ref}}} = \frac{1}{1 + G(s)} \tag{1}$$

Substituting (9.24) in (1) and setting $s = 0$ (for the steady state) yield

$$\Delta e_{ss} = \frac{(\Delta V_{\text{ref}})_{ss}}{1 + k} \quad \text{or} \quad 1 + k = \frac{(\Delta V_{\text{ref}})_{ss}}{\Delta e_{ss}} \tag{2}$$

The condition of the problem implies that the right side of (2) is not less than 100. Hence,

$$1 + k \geq 100$$

and $k \geq 99$.

9.16 Obtain the form of the dynamic response of the system of Fig. 9-3 to a step change in the reference input voltage.

From Fig. 9-3,

$$\Delta V_t(t) = \mathscr{L}^{-1}\left[\frac{G(s)}{1 + G(s)} \Delta V_{\text{ref}}(s)\right] \tag{1}$$

Where $G(s)$ is given in (9.24). The response of the system will depend on the characteristic roots of the equation

$$1 + G(s) = 0 \tag{2}$$

If the roots s_1, s_2, and s_3 are real and distinct, then the response will include the transient components $A_1 e^{s_1 t}$, $A_2 e^{s_2 t}$, and $A_3 e^{s_3 t}$. However, if (2) has a pair of complex conjugate roots $s_1, s_2 = \sigma \pm j\omega$, then the dynamic response will be of the form $Ae^{\sigma t}\sin(\omega t + \phi)$.

9.17 Assume that there are no changes occurring in the reference power setting of a turbine-governor system (that is, the system is operating in the steady state), and the frequency-power relationship of the turbine governor is that represented graphically in Fig. 9-7. Determine the regulation constant R.

In (9.26), we see that, with $\Delta P_{\text{ref}} = 0$, R is the negative of the slope of the f versus P_m curve, plotted in per-unit values. Hence, from Fig. 9-7,

$$R = \frac{\Delta f}{\Delta P_m} = \frac{0.01}{-0.2} = -0.05 \text{ pu}$$

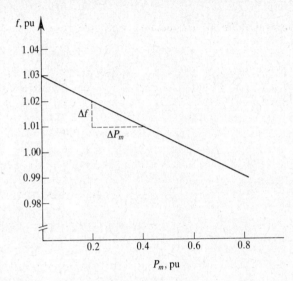

Fig. 9-7.

9.18 For a certain turbine-generator set, $R = 0.04$ pu, based on the generator rating of 100 MVA and 60 HZ. The generator frequency decreases by 0.02 Hz, and the system adjusts to steady-state operation. By how much does the turbine output power increase?

The per-unit frequency change is

$$\text{Per-unit } \Delta f = \frac{\Delta f}{f_{\text{base}}} = \frac{-0.02}{60} = -3.33 \times 10^{-4}\, \text{pu}$$

Then (9.26) yields

$$\text{Per-unit } \Delta P_m = -\frac{1}{0.04}(-3.33 \times 10^{-4}) = 8.33 \times 10^{-3}\, \text{pu}$$

The actual increase in output power is then

$$\Delta P_m = (8.33 \times 10^{-3})(100) = 0.833\, \text{MW}$$

9.19 An area includes two turbine-generator units, rated at 500 and 750 MVA and 60 Hz, for which $R_1 = 0.04$ pu and $R_2 = 0.05$ pu based on their respective ratings. Each unit carries a 300-MVA steady-state load. The load on the system suddenly increases by 250 MVA. (*a*) Calculate β on a 1000-MVA base. (*b*) Determine Δf on a 60-Hz base and in hertz.

(*a*) We can change the bases of the R values with the formula

$$R_{\text{new}} = R_{\text{old}} \frac{S_{\text{base(new)}}}{S_{\text{base(old)}}}$$

Thus

$$R_{1(\text{new})} = (0.04)\frac{1000}{500} = 0.08\, \text{pu}$$

and

$$R_{2(\text{new})} = (0.05)\frac{1000}{750} = 0.067\, \text{pu}$$

Now, from (9.30),

$$\beta = \frac{1}{R_1} + \frac{1}{R_2} = \frac{1}{0.08} + \frac{1}{0.067} = 27.5\, \text{pu}$$

(*b*) The per-unit increase in the load is $250/1000 = 0.25$ pu. From (9.29), with $\Delta P_{\text{ref(total)}} = 0$ for steady-state conditions,

$$\Delta f = \frac{-1}{\beta} \Delta P_m = -\frac{1}{27.5} 0.25 = -9.091 \times 10^{-3} \text{ pu}$$

Also,

$$\Delta f = -9.091 \times 10^{-3} \times 60 = -0.545 \text{ Hz}$$

9.20 For areas 1 and 2 in a 60-Hz power system, $\beta_1 = 400 \text{ MW/Hz}$ and $\beta_2 = 250 \text{ MW/Hz}$. The total power generated in each of these areas is, respectively, 1000 MW, and 750 MW. While each area is generating power at the steady state with $\Delta P_{\text{tie1}} = \Delta P_{\text{tie2}} = 0$, the load in area 1 suddenly increases by 50 MW. Determine the resulting Δf, (a) without LFC and (b) with LFC. Neglect all losses.

(a) From (9.29), since $\Delta P_{\text{ref(total)}} = 0$ without LFC,

$$50 = -(400 + 250) \Delta f$$

from which $\Delta f = -0.0769 \text{ Hz}$.

(b) With LFC, in the steady state, (9.27) implies that $\text{ACE}_1 = \text{ACE}_2 = 0$; otherwise, the LFC given by (9.27) would be changing the reference power settings of the governors on LFC. Also, the sum of the net tie-line flows, $\Delta P_{\text{tie1}} + \Delta P_{\text{tie2}}$, is zero (neglecting losses). So

$$\text{ACE}_1 + \text{ACE}_2 = 0 = (B_1 + B_2) \Delta f$$

and $\Delta f = 0$, since $B_1 + B_2 \neq 0$.

Supplementary Problems

9.21 A graph of fuel input versus power output for a certain plant is given in Fig. 9-8. Determine the fuel requirements at (a) 120 MW and (b) 560 MW output power.

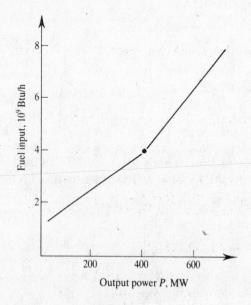

Fig. 9-8.

Ans. (a) $16.67 \times 10^6 \text{ Btu/MWh}$; (b) $10.71 \times 10^6 \text{ Btu/MWh}$

9.22 (a) For the plant of Problem 9.21, determine the fuel requirement at the maximum-efficiency operating point. (b) What is the power output at that point?

Ans. (a) $10 \times 10^6 \text{ Btu/MWh}$; (b) 400 MW

9.23   Assuming a fuel cost of $1.60 per million Btu for the plant of Problem 9.21, plot the incremental fuel cost versus output power.

   Ans.   Fig. 9-9

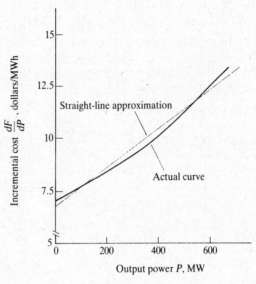

Fig. 9-9.

9.24   From the result of Problem 9.23, calculate the incremental fuel cost at the point at which the plant operates at maximum fuel efficiency.

   Ans. $16/MWh

9.25   Approximate the curve obtained in Problem 9.23 with a straight line, and obtain an equation for this line.

   Ans. $dF/dP = 0.0093P + 6.75$

9.26   The incremental fuel costs, in dollars per megawatthour, for two units in a plant are given by

$$\frac{dF_1}{dP_1} = 0.007P_1 + 7 \qquad \text{and} \qquad \frac{dF_2}{dP_2} = 0.009P_2 + 6$$

   During a 24-h period the load on each unit varies between 100 MW and 500 MW, whereas the plant output varies from 200 MW to 700 MW. (*a*) At what power level should unit 2 begin to take on all the additional load for the most economic operation of the plant? (*b*) What is the power output of the plant at this point? Neglect losses.

   Ans.   (*a*) 122.22 MW; (*b*) 222.22 MW

9.27   Referring to the result of Problem 9-5, plot a curve of incremental fuel cost versus plant output for minimum fuel cost for the plant of Problem 9.26.

9.28   When the plant of Problem 9.26 is delivering its maximum power output, how should this load be shared between the two units for minimum fuel cost?

   Ans.   331.25 MW; 368.75 MW

9.29   At what total output should the units in Problem 9.26 share the load equally?

   Ans.   1000 MW

9.30   Calculate the incremental fuel cost for the operating condition obtained in Problem 9.28.

   Ans. $9.319/MWh

9.31   At the maximum power output of the plant of Problem 9.26, determine the saving per hour in fuel cost for economic operation as compared to the situation in which unit 1 carries 400 MW.

   Ans. $37.81/h

9.32   For the system shown in Fig. 9-2, let $I_1 = 1.0\underline{/0°}$, $I_2 = 0.8\underline{/0°}$, $V_1 = 1.05\underline{/10°}$, and $V_2 = 1.07\underline{/15°}$, all per unit. The line impedances, again per unit, are $Z_1 = 0.05 + j0.20$, $Z_2 = 0.06 + j0.30$, and $Z_3 = 0.06 + j0.40$. Determine the system loss coefficients.

   Ans.   0.1029, 0.0561, and 0.1123 pu

9.33   Calculate the transmission loss for the system of Problem 9.32 with the loss formula (*9.6*). Verify the result by a direct calculation of the I^2R loss.

   Ans.   0.2828 pu

9.34   Express the result given by (*9.10*) in matrix form.

   Ans. $P_{\text{loss}} = \tilde{\mathbf{P}}\mathbf{B}\mathbf{P}$, where $\tilde{\mathbf{P}} = \mathbf{P}$ transposed

9.35   Calculate the penalty factors for the two plants operating as in Problem 9.32.

   Ans.   1.35; 1.32

9.36   For the system of Problem 9.32, the incremental fuel costs for the two plants are given by

$$\frac{dF_1}{dP_1} = 0.01P_1 + 10 \quad \text{and} \quad \frac{dF_2}{dP_2} = 0.02P_2 + 10$$

   The system operates at $\lambda = \$11.6/\text{MWh}$ for minimum fuel cost. Determine the power generated at each plant.

   Ans.   160 MW; 80 MW

9.37   Calculate the efficiency of the plant of Problem 9.32.

   Ans.   71.72 percent

9.38   The incremental fuel costs for two units in a plant are

$$\frac{dF_1}{dP_1} = 0.20P_1 + 4.0 \quad \text{and} \quad \frac{dF_2}{dP_2} = 0.25P_2 + 3.0$$

   The minimum and maximum loads on each unit are 20 MW and 125 MW, whereas those on the plant are 40 MW and 250 MW. At what plant output should the first unit begin to share the load for most economic operation of the plant?

   Ans.   40 MW

9.39   Determine the incremental fuel cost for the given operation of the plant of Problem 9.38.

   Ans. $8.0/MWh

9.40   For the system of Fig. 9-3, calculate the percent steady-state error at an open-loop gain of 90.

   Ans.   1.1 percent

9.41   A 500-MW, 60-Hz generator has a regulation constant of 0.05. Calculate the increase in the input power resulting from a 0.2-Hz drop in the frequency, with no change in the reference input to the system.

Ans.   20 MW

9.42   For the system of Problem 9.19, determine the per-unit increase in the input to each generator that results from the load increase.

Ans.   0.18125 pu; 0.21642 pu

9.43   The operating costs for two units supplying power to a system are in dollars per hour,

$$F_1 = 10P_1 + 8000P_1^2 \quad \text{and} \quad F_2 = 8P_2 + 9000P_2^2$$

where P_1 and P_2 are in kilowatts. For a total output of 800 MW, calculate the output from each unit such that the total operating cost is minimal.

Ans.   365 MW; 435 MW

9.44   Determine the minimum hourly operating cost for the system of Problem 9.43 with a 600-MW load on the system.

Ans. $6900/h

9.45   Calculate the incremental operating cost that minimizes the total operating cost for the system of Problem 9.43 when the total load on the system is 1300 MW.

Ans. $20/MWh

Chapter 10

Power System Stability

By the *stability* of a power system we mean the ability of the system to remain in operating equilibrium, or synchronism, while disturbances occur on the system. Three types of stability are of concern: steady-state, dynamic, and transient stability.

Steady-state stability relates to the response of a synchronous machine to a gradually increasing load.

Dynamic stability involves the response to small disturbances that occur on the system, producing oscillations. If these oscillations are of successively smaller amplitudes, the system is considered dynamically stable. If the oscillations grow in amplitude, the system is dynamically unstable. The source of this type of instability is usually an interaction between control systems. The system's response to the disturbance may not become apparent for some 10 to 30 s.

Transient stability involves the response to large disturbances, which may cause rather large changes in rotor speeds, power angles, and power transfers. The system's response to such a disturbance is usually evident within 1 s.

10.1 INERTIA CONSTANT AND SWING EQUATION

The angular momentum and inertia constant play an important role in determining the stability of a synchronous machine. The *per-unit inertia constant H* is defined as the kinetic energy stored in the rotating parts of the machine *at synchronous speed* per unit megavoltampere (MVA) rating of the machine. Thus, if G is the MVA rating of the machine, then

$$GH = \tfrac{1}{2}J\omega_s^2 \qquad (10.1)$$

where J is the polar moment of inertia of all rotating parts in kilogram-(meters-squared), and ω_s is the angular synchronous velocity in electrical radians per second. If M is the corresponding angular momentum, then

$$M = J\omega_s \qquad (10.2)$$

Since $\omega_s = 360f$ electrical degrees per second, (10.1) and (10.2) yield

$$GH = \tfrac{1}{2}M\omega_s = \tfrac{1}{2}M(360)f$$

or

$$M = \frac{GH}{180f} \quad \text{MJ} \cdot \text{s/electrical degree} \qquad (10.3)$$

where f is the frequency of rotation.

Consider a synchronous generator developing an electromagnetic torque T_e (and a corresponding electromagnetic power P_e) while operating at the synchronous speed ω_s. If the input torque provided by the prime mover at the generator shaft is T_i, then under steady-state conditions (with no disturbance) we have

$$T_e = T_i$$

or

$$T_e\omega_s = T_i\omega_s$$

and

$$T_i\omega_s - T_e\omega_s = P_i - P_e = 0 \qquad (10.4)$$

If a departure from steady state occurs, such as a change in load or a fault, the "power in" P_i no longer equals the "power out" P_{out}, and the left side of (10.4) is not zero. Instead, an accelerating torque comes into play. If P_a is the corresponding accelerating (or decelerating) power, then

$$P_a = P_i - P_e = M\frac{d^2\theta}{dt^2} \qquad (10.5)$$

146

where M has been defined in (10.3), P_a is in megawatts, and θ is the angular position of the rotor. Further, in the steady state,

$$\frac{d\theta}{dt} = \omega_s$$

so

$$\theta = \omega_s t + \delta \qquad\qquad (10.6)$$

where the constant of integration δ is called the *power angle* of the synchronous machine. Substituting (10.6) in (10.5) yields

$$M\frac{d^2\delta}{dt^2} = P_i - P_e = P_a \qquad\qquad (10.7)$$

which is known as the *swing equation*. If we combine (10.3) and (10.7) and divide by G, we obtain the per-unit swing equation as

$$\frac{H}{180f}\frac{d^2\delta}{dt^2} = P_i - P_e = P_a \qquad \text{per unit} \qquad\qquad (10.8)$$

The swing equation contains information regarding the machine dynamics and stability. However, it is important to realize that we made two basic assumptions in deriving it: (1) In (10.2) we took M to be constant, although, strictly speaking, this is not so; (2) the damping term proportional to $d\delta/dt$ has been neglected.

10.2 H CONSTANT ON A COMMON MVA BASE

An inertia constant H_{mach} based on a machine's own MVA rating may be converted to a value H_{syst} relative to the system base S_{syst} with the formula

$$H_{\text{syst}} = H_{\text{mach}}\frac{S_{\text{mach}}}{S_{\text{syst}}} \qquad\qquad (10.9)$$

A convenient system base value is 100 MVA.

The moment of inertia of a synchronous machine is given by $WR^2/32.2$ slug-(feet-squared), where W is the weight of the rotating part of the machine in pounds, and R is its radius of gyration in feet. Machinery manufacturers generally supply the value of WR^2 for their machines.

10.3 EQUAL-AREA CRITERION

Consider δ in the swing equation (10.7), which describes the motion, or *swing*, of the rotor. As is shown in Fig. 10-1, in an unstable system, δ increases indefinitely with time and the machine loses synchronism. In a stable system, δ undergoes oscillations which eventually die out. From the figure it is clear that, for a system to be stable, it must be that $d\delta/dt = 0$ at some instant. This criterion (that $d\delta/dt$ be zero) can be obtained simply from (10.7). Furthermore, if we assume that H is constant and that damping is negligible and we ignore the control system, then we have

$$2\frac{d\delta}{dt}\frac{d^2\delta}{dt^2} = \frac{2P_a}{M}\frac{d\delta}{dt}$$

which, upon integration, gives

$$\left(\frac{d\delta}{dt}\right)^2 = \frac{2}{M}\int_{\delta_0}^{\delta} P_a\, d\delta$$

so that

$$\frac{d\delta}{dt} = \sqrt{\frac{2}{M} \int_{\delta_0}^{\delta} P_a \, d\delta}$$

where δ_0 is the initial power angle before the rotor begins to swing because of a disturbance. The stability criterion $d\delta/dt = 0$ (at some moment) implies that

$$\int_{\delta_0}^{\delta} P_a \, d\delta = 0 \qquad (10.10)$$

This condition requires that, for stability, the area under the graph of accelerating power P_a versus δ (Fig. 10-2) must be zero for some value of δ; that is, the positive (or accelerating) area under the graph must be equal to the negative (or decelerating) area. This criterion is therefore known as the *equal-area criterion* for stability.

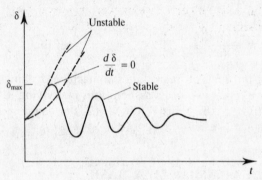

Fig. 10-1.

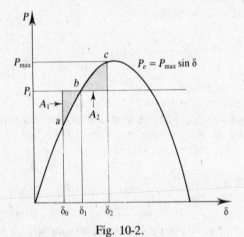

Fig. 10-2.

The equal-area criterion requires that, for stability,

$$\text{Area } A_1 = \text{area } A_2$$

or

$$\int_{\delta_0}^{\delta_1} (P_i - P_{max} \sin \delta) \, d\delta = \int_{\delta_1}^{\delta_2} (P_{max} \sin \delta - P_i) \, d\delta$$

or, after the integrations are performed,

$$P_i(\delta_1 - \delta_0) + P_{max}(\cos \delta_1 - \cos \delta_0) = P_i(\delta_1 - \delta_2) + P_{max}(\cos \delta_1 - \cos \delta_2) \qquad (10.11)$$

But because

$$P_i = P_{max} \sin \delta_1$$

(*10.11*) becomes

$$(\delta_2 - \delta_0) \sin \delta_1 + \cos \delta_2 - \cos \delta_0 = 0 \qquad (10.12)$$

If we know δ_0 and δ_1, we can solve (*10.12*) for δ_2.

10.4 CRITICAL CLEARING ANGLE

If a disturbance (or fault) occurs in a system, δ begins to increase under the influence of positive accelerating power, and the system will become unstable if δ becomes very large. There is a critical angle within which the fault must be cleared if the system is to remain stable and the equal-area criterion is to be satisfied. This angle is known as the *critical clearing angle* δ_c. As an example, consider a system that normally operates along curve A in Fig. 10-3. If a three-phase short circuit occurs across the line, its curve of power versus power angle will correspond to the horizontal axis. For stability, the critical clearing angle must be such that area A_1 = area A_2.

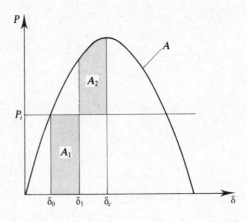

Fig. 10-3.

In Fig. 10-4 we show a power-angle curve A before a fault, B during the fault, and C after the fault such that $A = P_{\max} \sin \delta$, $B = k_1 A$, and $C = k_2 A$, with $k_1 < k_2$. For stability, we must have area A_1 = area A_2. Based on Fig. 10-4, this condition yields

$$(\delta_m - \delta_0)P_i = \int_{\delta_0}^{\delta_c} B \, d\delta + \int_{\delta_c}^{\delta_m} C \, d\delta \qquad (10.13)$$

Substituting for B and C in (*10.13*), with $P_i = P_{\max} \sin \delta_0$, eventually yields

$$\cos \delta_c = \frac{1}{k_2 - k_1}[(\delta_m - \delta_0) \sin \delta_0 - k_1 \cos \delta_0 + k_2 \cos \delta_m] \qquad (10.14)$$

From Fig. 10-4, we have

$$P_i = P_m \sin \delta_0 = k_2 P_m \sin \delta_m = k_2 P_m \sin (\pi - \delta_m) \qquad (10.15)$$

Hence, from (*10.15*),

$$\sin \delta_0 = k_2 \sin (\pi - \delta_m) \qquad (10.16)$$

With k_1, k_2, and δ_0 specified, the critical clearing angle may be obtained from (*10.14*) and (*10.16*).

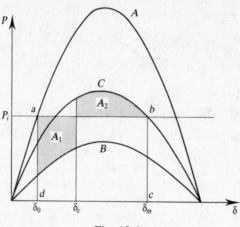

Fig. 10-4.

10.5 A TWO-MACHINE SYSTEM

The swing equation (10.7) may be written for two machines as

$$M_1 \frac{d^2\delta_1}{dt^2} = P_{i1} - P_{e1} \tag{10.17}$$

$$M_2 \frac{d^2\delta_2}{dt^2} = P_{i2} - P_{e2} \tag{10.18}$$

where the subscripts 1 and 2 correspond to machines 1 and 2, respectively. If we denote the relative angle between the two rotor axes by δ, such that $\delta = \delta_1 - \delta_2$, then (10.17) and (10.18) may be combined and simplified to

$$M \frac{d^2\delta}{dt^2} = P_i - P_e \qquad \text{for two machines} \tag{10.19}$$

where

$$M = \frac{M_1 M_2}{M_1 + M_2} \qquad P_i = \frac{M_2 P_{i1} - M_1 P_{i2}}{M_1 + M_2} \qquad P_e = \frac{M_2 P_{e1} - M_1 P_{e2}}{M_1 + M_2}$$

10.6 STEP-BY-STEP SOLUTION

The swing equation may be solved iteratively with the step-by-step procedure shown in Fig. 10-5. In the solution, it is assumed that the accelerating power P_a and the relative rotor angular velocity ω_r are constant within each of a succession of intervals (top and middle, Fig. 10-5); their values are used to find the change in δ during each interval.

To begin the iterations, we need $P_a(0+)$, which we evaluate as

$$P_a(0+) = P_i - P_e(0+) \tag{10.20}$$

Then the swing equation may be written

$$\frac{d^2\delta}{dt^2} = \alpha(0+) = \frac{P_a(0+)}{M} \tag{10.21}$$

and the change in ω_r is given (Fig. 10-5) by

$$\Delta\omega_r = \alpha(0+)\,\Delta t \tag{10.22}$$

Then

$$\omega_r = \omega_0 + \Delta\omega_r = \omega_0 + \alpha(0+)\,\Delta t \tag{10.23}$$

Similarly, the change in the power angle for the first interval is

$$\Delta\delta_1 = \Delta\omega_r\,\Delta t \tag{10.24}$$

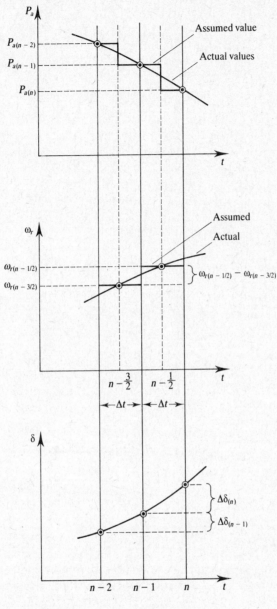

Fig. 10-5.

and so
$$\delta_1 = \delta_0 + \Delta\delta_1 = \delta_0 + \alpha(0+)(\Delta t)^2 \tag{10.25}$$

Evaluation of P_a

If there is no discontinuity in the swing curve during an iteration interval, then $P_a(0+)$ is equal to half of P_a immediately after the fault. (A swing curve is a curve of power versus δ.) If there is a discontinuity at the beginning of the ith interval, then

$$P_{a(i-1)} = \tfrac{1}{2}(P_{a(i-1)-} + P_{a(i-1)+}) \tag{10.26}$$

where $P_{a(i-1)-}$ and $P_{a(i-1)+}$ are, respectively, the accelerating power immediately before and immediately after the fault is cleared.

If the discontinuity occurs at the middle of an interval, then for that interval

$$P_a = P_i - \text{output during the fault} \tag{10.27}$$

For this case, at the beginning of the interval immediately following the clearing of the fault, P_a is given by

$$P_a = P_i - \text{output after the fault is cleared} \tag{10.28}$$

Finally, if the discontinuity occurs neither at the beginning nor at the middle of an interval, P_a may still be evaluated from (10.26) through (10.28).

Algorithm for the Iterations

Returning now to (10.25), we see that δ_1 gives us one point on the swing curve. The algorithm for the iterative process is as follows:

$$P_{a(n-1)} = P_i - P_{e(n-1)} \tag{10.29}$$

$$P_{e(n-1)} = \frac{|E||V|}{X} \sin \delta_{(n-1)} \tag{10.30}$$

$$\alpha_{(n-1)} = \frac{P_{a(n-1)}}{M} \tag{10.31}$$

$$\Delta\omega_{r(n)} = \alpha_{(n-1)} \Delta t \tag{10.32}$$

$$\omega_{r(n)} = \omega_{r(n-1)} + \alpha_{(n-1)} \Delta t \tag{10.33}$$

$$\Delta\delta_{(n)} = \Delta\delta_{(n-1)} + \frac{P_{a(n-1)}}{M} (\Delta t)^2 \tag{10.34}$$

$$\delta_{(n)} = \delta_{(n-1)} + \Delta\delta_{(n)} \tag{10.35}$$

The use of this algorithm in conjunction with the equal-area criterion provides the critical clearing angle and the corresponding critical clearing time.

Solved Problems

10.1 The inertia constant H for a 60-Hz, 100-MVA hydroelectric generator is 4.0 MJ/MVA. How much kinetic energy is stored in the rotor at synchronous speed? If the input to the generator is suddenly increased by 20 MVA, what acceleration is imparted to the rotor?

The energy stored in the rotor at synchronous speed is given by (10.1) and is

$$GH = 100 \times 4 = 400 \text{ MJ}$$

The rotor acceleration $d^2\delta/dt^2$ is given by (10.7) with $P_a = 20$ MVA of accelerating power and with M as determined from (10.3). Thus, (10.3) yields

$$M = \frac{GH}{180f} = \frac{400}{180 \times 60} = \frac{1}{27}$$

and (10.7) becomes

$$\frac{1}{27} \frac{d^2\delta}{dt^2} = 20$$

so $d^2\delta/dt^2 = 20 \times 27 = 540°/\text{s}^2$.

10.2 In Section 10.2 we noted that machinery manufacturers generally supply the value of WR^2.

Derive a relationship between H and WR^2 for a machine whose rating is S_{mach} MVA.

The kinetic energy of rotation of the rotor at synchronous speed is

$$KE = \frac{1}{2}\frac{WR^2}{32.2}\left(\frac{2\pi n}{60}\right)^2 \quad \text{(in foot-pounds)}$$

where n is the rotor speed in revolutions per minute. Since $550\,\text{ft}\cdot\text{lb/s} = 746\,\text{W}$, $1\,\text{ft}\cdot\text{lb} = 746/550\,\text{J}$. Converting foot-pounds to megajoules and dividing the last equation by the machine rating in megavoltamperes, we obtain

$$H = \frac{\left(\dfrac{746}{550}\times 10^{-6}\right)\left(\dfrac{1}{2}\dfrac{WR^2}{32.2}\right)\left(\dfrac{2\pi n}{60}\right)^2}{S_{mach}}$$

$$= \frac{2.31\times 10^{-10}WR^2 n^2}{S_{mach}} \tag{1}$$

10.3 A 1500-MVA, 1800-rev/min synchronous generator has $WR^2 = 6\times 10^6\,\text{lb}\cdot\text{ft}^2$. Find the inertia constant H of the machine relative to a 100-MVA base.

From (1) of Problem 10.2,

$$H = \frac{(2.31\times 10^{-10})(6\times 10^6)(1800)^2}{1500} = 2.994\,\text{MJ/MVA}$$

Relative to a 100-MVA base, then,

$$H = 2.994\times\frac{1500}{100} = 44.91\,\text{MJ/MVA}$$

10.4 A 500-MVA synchronous machine has $H_1 = 4.6\,\text{MJ/MVA}$, and a 1500-MVA machine has $H_2 = 3.0\,\text{MJ/MVA}$. The two machines operate in parallel in a power station. What is the equivalent H constant for the two, relative to a 100-MVA base?

The total kinetic energy of the two machines is

$$KE = 4.6\times 600 + 3\times 1500 = 6800\,\text{MJ}$$

Thus, the equivalent H relative to a 100-MVA base is

$$H = \frac{6800}{100} = 68\,\text{MJ/MVA}$$

10.5 For a certain lagging-power-factor load, the sending-end and receiving-end voltages of a short transmission line of impedance $R + jX$ are equal. Determine the ratio X/R so that maximum power is transmitted over the line under steady-state conditions.

From the phasor diagram of Fig. 10-6, we may write

$$V_S = V_R + I(\cos\phi - j\sin\phi)(R + jX)$$
$$= (V_R + IR\cos\phi + IX\sin\phi) + j(IX\cos\phi - IR\sin\phi)$$
$$R(V_S\cos\delta) = (V_R + IR\cos\phi + IX\sin\phi)R$$
$$X(V_S\sin\delta) = (IX\cos\phi - IR\sin\phi)X$$

Combining these equations and letting $Z^2 = R^2 + X^2$, we get

$$V_S(R\cos\delta + X\sin\delta) = RV_R + IZ^2\cos\phi$$

or

$$I\cos\phi = \frac{V_S}{Z^2}(R\cos\delta + X\sin\delta) - \frac{RV_R}{Z^2}$$

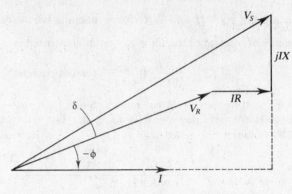

Fig. 10-6.

Hence, we have

$$P_R = V_R I \cos \phi = \frac{V_R V_S}{Z^2}(R \cos \delta + X \sin \delta) - \frac{R V_R^2}{Z^2} \qquad (1)$$

Now let $\tan \beta = X/R$; then (1) becomes

$$P_R = \frac{V_R V_S}{Z} \cos (\beta - \delta) - \frac{R V_R^2}{Z^2}$$

For maximum power $\beta = \delta$, and so

$$P_{R(\text{max})} = \frac{V_R V_S}{\sqrt{R^2 + X^2}} - \frac{R V_R^2}{R^2 + X^2} \qquad (2)$$

and

$$\frac{dP_{R(\text{max})}}{dX} = 0$$

Thus

$$\left(\frac{V_S}{V_R}\right)^2 (R^2 + X^2) = 4R^2$$

and since $V_S = V_R$, we have $X/R = \sqrt{3}$.

10.6 The sending-end and receiving-end voltages of a transmission line at a 100-MW load are equal at 115 kV. The per-phase line impedance is $(4 + j7)\ \Omega$. Calculate the maximum steady-state power that can be transmitted over the line.

Since $V_R = V_S = 115{,}000/\sqrt{3} = 66{,}400$, we have, from (2) of Problem 10.5,

$$P_{R(\text{max})} = \frac{V_R V_S}{\sqrt{R^2 + X^2}} - \frac{R V_R^2}{R^2 + X^2}$$

$$= \left[\frac{(66.4)^2}{\sqrt{4^2 + 7^2}} - \frac{4(66.4)^2}{4^2 + 7^2}\right] 10^6 = 275.5\ \text{MW/phase}$$

$$= 826.5\ \text{MW total}$$

10.7 A synchronous generator, capable of developing 500 MW of power, operates at a power angle of 8°. By how much can the input shaft power be increased suddenly without loss of stability?

Initially, at $\delta_0 = 8°$, the electromagnetic power being developed is

$$P_{e0} = P_{\text{max}} \sin \delta_0 = 500 \sin 8° = 69.6\ \text{MW}$$

Let δ_m (Fig. 10-7) be the power angle to which the rotor can swing before losing synchronism. Then the equal-area criterion requires that (10.12) be satisfied (with δ_m replacing δ_2). From Fig. 10-7,

$\delta_m = \pi - \delta_1$, so ($10.12$) yields

$$(\pi - \delta_1 - \delta_0) \sin \delta_1 + \cos (\pi - \delta_1) - \cos \delta_0 = 0$$

or
$$(\pi - \delta_1 - \delta_0) \sin \delta_1 - \cos \delta_1 - \cos \delta_0 = 0 \qquad (1)$$

Substituting $\delta_0 = 8° = 0.13885$ rad in (1) gives

$$(3 - \delta_1) \sin \delta_1 - \cos \delta_1 - 0.99 = 0$$

This yields $\delta_1 = 50°$, for which the corresponding after-the-fault electromagnetic power is

$$P_{ef} = P_{max} \sin \delta_1 = 500 \sin 50° = 383.02 \text{ MW}$$

The initial power developed by the machine was 69.6 MW. Hence, without loss of stability, the system can accommodate a sudden increase of

$$P_{ef} - P_{e0} = 383.02 - 69.6 = 313.42 \text{ MW}$$

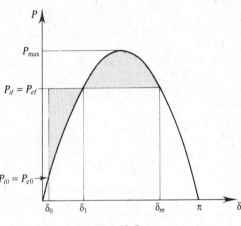

Fig. 10-7.

10.8 Determine the maximum additional load that could suddenly be taken on by the transmission line of Problem 10.6 without losing stability.

If we neglect the resistance, then the initial (maximum) power P_0 is

$$P_0 = \frac{V_S V_R}{X} \sin \delta_0 = P_{R(max)} \sin \delta_0$$

From (1) of Problem 10.7,

$$(\pi - \delta_1 - \delta_0) \sin \delta_1 - \cos \delta_1 - \cos \delta_0 = 0 \qquad (1)$$

We have
$$P_0 = \tfrac{1}{3}(100) = 33.33 \text{ MW}$$

and
$$P_{R(max)} = \frac{1}{3} \frac{115^2}{7} 10^6 = 629.76 \text{ MW}$$

so
$$\delta_0 = \sin^{-1} \frac{33.33}{629.76} = 3° = 0.052 \text{ rad}$$

Then (1) becomes

$$(\pi - \delta_1 - 0.052) \sin \delta_1 - \cos \delta_1 - \cos 3° = 0$$

which yields $\delta_1 = 47.8°$. Hence the system will remain stable for an increase in load of up to

$$P_{R(max)} \sin \delta_1 - P_0 = 629.76 \sin 47.8° - 33.33 = 433.2 \text{ MW/phase}$$

$$= 1299.6 \text{ MW total}$$

10.9   A synchronous generator is operating at an infinite bus and supplying 0.45 pu of its maximum power capacity. A fault occurs, and the reactance between the generator and the line becomes four times its value before the fault. The maximum power that can be delivered after the fault is cleared is 70 percent of the original maximum value. Determine the critical clearing angle.

Let

$$x_1 = \frac{P_{\max} \text{ during the fault}}{P_{\max} \text{ before the fault}}$$

$$x_2 = \frac{P_{\max} \text{ after the fault}}{P_{\max} \text{ before the fault}}$$

δ_0 = power angle at the time of the fault

δ_c = power angle when fault is cleared

δ_m = maximum angle of swing

Then the equal-area criterion, $A_1 = A_2$ in Fig. 10-8, gives us

$$P_s(\delta_c - \delta_0) - \int_{\delta_0}^{\delta_c} x_1 P_{\max} \sin \delta \, d\delta = \int_{\delta_c}^{\delta_m} x_2 P_{\max} \sin \delta \, d\delta - P_s(\delta_m - \delta_c)$$

Hence, $$\cos \delta_c = \frac{1}{x_2 - x_1} \left[\frac{P_s}{P_{\max}} (\delta_m - \delta_0) + x_2 \cos \delta_m - x_1 \cos \delta_0 \right] \qquad (1)$$

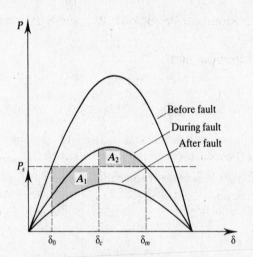

Fig. 10-8.

Initially, the generator is supplying 0.45 pu of $P_{\max}$. Thus,

$$P_s = 0.45 P_{\max} = P_{\max} \sin \delta_0$$

from which $\delta_0 = \sin^{-1} 0.45 = 26.74°$. Now $P_{\max} = EV/X$. When the fault occurs, X becomes $4X$, so that

$$x_1 P_{\max} \sin \delta_m = \frac{EV}{4X} \sin \delta_m = \tfrac{1}{4} P_{\max} \sin \delta_m$$

so that $x_1 = 0.25$.

After the fault, with $x_2 = 0.70$, we have

$$P_s = x_2 P_{\max} \sin \delta_m'$$

from which

$$\delta_m' = \sin^{-1} \frac{P_s}{x_2 P_{\max}} = \sin^{-1} \frac{0.45 P_{\max}}{0.70 P_{\max}} = 40°$$

Then $\delta_m = 90° + \delta'_m = 130°$ (see Fig. 10-8), and

$$\delta_m - \delta_0 = 130° - 26.74° = 103.26° \qquad \text{or} \qquad 1.8019\,\text{rad}$$

Hence, from (1),

$$\cos\delta_c = \frac{1}{0.70 - 0.25}[0.45(1.8019) + 0.70\cos 130° - 0.25\cos 26.74°] = 0.3059$$

so that $\delta_c = \cos^{-1} 0.3059 = 72.2°$.

10.10 A 100-MVA, two-pole, 60-Hz generator has a moment of inertia of $50 \times 10^3\,\text{kg}\cdot\text{m}^2$. What is the energy stored in the rotor at the rated speed? What is the corresponding angular momentum? Determine the inertia constant H.

The stored energy is

$$\text{KE(stored)} = \frac{1}{2}J\omega_m^2 = \frac{1}{2}(50 \times 10^3)\left(\frac{2\pi \times 3600}{60}\right)^2 = 3553\,\text{MJ}$$

Then
$$H = \frac{\text{KE(stored)}}{\text{MVA}} = \frac{3553}{100} = 35.53\,\text{MJ/MVA}$$

$$M = \frac{GH}{180f} = \frac{(100)(35.53)}{(180)(60)} = 0.329\,\text{MJ}\cdot\text{rad/s}$$

10.11 The input to the generator of Problem 10.10 is suddenly increased by 25 MW. Determine the rotor acceleration.

From Problems 10.10 and 10.1,

$$0.329\ddot{\delta} = 25$$

Thus,
$$\ddot{\delta} = \frac{25}{0.329} = 76°/\text{s}^2$$

10.12 Assuming the acceleration calculated in Problem 10.11 remains constant for twelve cycles, calculate the change in the power angle and the speed that occurs during those twelve cycles.

Twelve cycles are equivalent to $12/60 = 0.2\,\text{s}$. During that time, δ changes by $\frac{1}{2}(471.25)(0.2)^2 = 9.425$ electrical degrees. Now
$$\ddot{\delta} = 60 \times \frac{471.25}{360} = 78.5\,\text{rev/min/s}$$

so the rotor speed at the end of the twelve cycles is $3600 + 78.5 = 3678.5\,\text{rev/min}$.

10.13 A 60-Hz generator, connected directly to an infinite bus operating at a voltage of $1\underline{/0°}$ pu, has a synchronous reactance of 1.35 pu. The generator no-load voltage is 1.1 pu, and its inertia constant H is 4 MJ/MVA. The generator is suddenly loaded to 60 percent of its maximum power limit; determine the frequency of the resulting natural oscillations of the generator rotor.

We find δ_0 using $\sin\delta_0 = P_e/P_i = 0.6$, which gives $\delta_0 = 36.87°$. Then

$$\left.\frac{\partial P_e}{\partial\delta}\right|_{36.87} = \frac{1.1 \times 1}{1.35}\cos 36.87 = 0.6518\,\text{pu/rad}$$

Also, we have

$$M = \frac{H}{\pi f} = \frac{4}{\pi \times 60}\,\text{pu s}^2/\text{rad}$$

$$\text{Frequency of oscillation} = \sqrt{\frac{(\partial P_e / \partial \delta)_{36.87}}{M}}$$

$$= \sqrt{\frac{\pi \times 60 \times 0.6518}{4}} = 5.5 \,\text{rad/s} = 0.882 \,\text{Hz}$$

10.14 Derive (*10.19*).

Since $\delta = \delta_1 - \delta_2$,

$$\ddot{\delta} = \ddot{\delta}_1 - \ddot{\delta}_2 \tag{1}$$

From (*1*), (*10.17*), and (*10.18*), we have

$$\ddot{\delta}_1 - \ddot{\delta}_2 = \frac{1}{M_1}(P_{i1} - P_{e1}) - \frac{1}{M_2}(P_{i2} - P_{e2}) = \ddot{\delta} \tag{2}$$

Multiplying both sides of (*2*) by $M_1 M_2 / (M_1 + M_2)$ yields

$$\frac{M_1 M_2}{M_1 + M_2} \ddot{\delta} = \frac{1}{M_1 + M_2}[(M_2 P_{i1} - M_1 P_{i2}) - (M_2 P_{e1} - M_1 P_{e2})]$$

$$= \frac{M_2 P_{i1} - M_1 P_{i2}}{M_1 + M_2} - \frac{M_2 P_{e1} - M_1 P_{e2}}{M_1 + M_2}$$

or

$$M\ddot{\delta} = P_i - P_e$$

which is the same as (*10.19*).

10.15 The kinetic energy stored in the rotor of a 50-MVA, six-pole, 60-Hz synchronous machine is 200 MJ. The input to the machine is 25 MW at a developed power of 22.5 MW. Calculate the accelerating power and the acceleration.

The accelerating power is

$$P_a = P_i - P_e = 25 - 22.5 = 2.5 \,\text{MW}$$

Now, also,

$$H = \frac{\text{KE(stored)}}{\text{machine rating in MVA}} = \frac{200}{50} = 4$$

and, from (*10.3*),

$$M = \frac{GH}{180f} = \frac{50 \times 4}{180 \times 60} = 0.0185 \,\text{MJ} \cdot \text{s/degree}$$

$$= 1.06 \,\text{MJ} \cdot \text{s/rad}$$

Finally, from (*10.7*),

$$\ddot{\delta} = \frac{2.5}{1.06} = 2.356 \,\text{rad/s}^2$$

10.16 If the acceleration of the machine of Problem 10.15 remains constant for ten cycles, what is the power angle at the end of the ten cycles?

From Problem 10-15, $\ddot{\delta} = 2.356$. Integration with respect to t yields

$$\dot{\delta} = 2.356t + C_1$$

Since $\dot{\delta} = 0$ at $t = 0$, $C_1 = 0$. A second integration now gives

$$\delta = 1.178t^2 + C_2$$

At $t = 0$, let $\delta = \delta_0$ (the initial power angle). Then

$$\delta = 1.178t^2 + \delta_0$$

At 60 Hz, the time required for ten cycles is $t = \frac{1}{6}$ s. For this value of t,

$$\delta = 1.178(\tfrac{1}{6})^2 + \delta_0 = (0.0327 + \delta_0)\,\text{rad}$$

10.17 The generator of Problem 10.15 has an internal voltage of 1.2 pu and is connected to an infinite bus operating at a voltage of 1.0 pu through a 0.3-pu reactance. A three-phase short circuit occurs on the line. Subsequently, circuit breakers operate and the reactance between the generator and the bus becomes 0.4 pu. Calculate the critical clearing angle.

Before the fault,

$$P_{\text{max}} = \frac{1.2 \times 1.0}{0.3} = 4.0\,\text{pu}$$

During the fault,

$$P_{\text{max2}} = 0$$

and $k_1 = 0$ for use in (10.14). After the fault is cleared,

$$P_{\text{max3}} = \frac{1.2 \times 1.0}{0.4} = 3.0\,\text{pu}$$

and $k_2 = 3.0/4.0 = 0.75$ for use in (10.14).

The initial power angle δ_0 is given by $4 \sin \delta_0 = 1.0$, from which $\delta_0 = 0.2527\,\text{rad}$. Define $\delta_m' = \pi - \delta_m$ (see Fig. 10-4). The angle δ_m in (10.14) is obtained from

$$\sin \delta_m' = \frac{1}{3.0} \qquad \text{and} \qquad \delta_m = \pi - \delta_m'$$

from which $\delta_m = 2.8\,\text{rad}$. Substituting k_1, k_2, δ_0 and δ_m in (10.14) yields

$$\cos \delta_c = \frac{1}{0.75}[(2.8 - 0.2527)0.25 - 0 + 0.75 \cos 2.8] = -0.093$$

from which $\delta_c = 95.34°$.

10.18 Using the step-by-step algorithm, plot the swing curve for the machine of Problem 10.17.

The per-unit value of the angular momentum, based on the machine rating, is

$$M = \frac{1.0 \times 4}{180 \times 60} = 3.7 \times 10^{-4}\,\text{pu}$$

From (10.26), we have

$$P_a(0+) = \frac{1.0 - 0.0}{2} = 0.5$$

From (10.21),

$$\alpha(0+) = \frac{0.5}{3.7 \times 10^4} = 1351°/\text{s}$$

From (10.22) with $\Delta t = 0.05$ s,

$$\Delta\omega_{r(1)} = 1351 \times 0.05 = 67.55°/\text{s}$$

From (10.23),

$$\omega_{r(1)} = 0 + 67.55 = 67.55°/\text{s}$$

From (10.24),

$$\Delta\delta_{(1)} = 67.55 \times 0.05 = 3.3775°$$

Finally, from (10.25), with $\delta_0 = 14.4775°$ as determined in Problem 10.17,

$$\delta_{(1)} = 14.4775 + 3.3775 = 17.855°$$

For the second interval, (10.29) and (10.31) to (10.35) give us

$$P_{a(1)} = 1.0 - 0.0 = 1.0$$

$$\alpha_{(1)} = \frac{1.0}{3.7 \times 10^4} = 2702°/s$$

$$\Delta\omega_{r(2)} = 2702 \times 0.05 = 135.1°$$

$$\omega_{r(2)} = \omega_{r(1)} + \Delta\omega_{r(2)} = 67.55 + 135.1 = 202.65°/s$$

$$\Delta\delta_{(2)} = \omega_{r(2)}\Delta t = 202.65 \times 0.05 = 10.1325°$$

$$\delta_{(2)} = \delta_{(1)} + \Delta\delta_{(2)} = 17.855 + 10.1325 = 27.9875°$$

Since α and $\Delta\omega_r$ do not change during succeeding intervals, we have

$$\omega_{r(3)} = \omega_{r(2)} + \Delta\omega_{r(3)} = 337.75°/s$$

$$\Delta\delta_{(3)} = \omega_{r(3)}\Delta t = 337.75 \times 0.05 = 16.8875°$$

$$\delta_{(3)} = \delta_{(2)} + \Delta\delta_{(3)} = 44.875°$$

and so on. In this way we obtain the following table of values, from which Fig. 10-9 is plotted:

t, s	δ, degrees
0.0	14.48
0.05	17.85
0.10	27.99
0.15	44.88
0.20	68.52
0.25	98.92

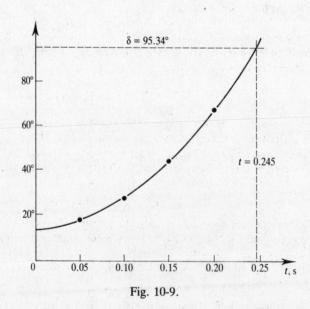

Fig. 10-9.

10.19 From the results of Problems 10.17 and 10.18, find the critical clearing time in cycles for an appropriately set circuit breaker.

From Problem 10.17, $\delta_c = 95.34°$. For this critical clearing angle, Fig. 10-9 gives $t = 0.245$ s. Hence the fault must be cleared within $60 \times 0.245 = 14.7$ cycles.

Supplementary Problems

10.20 The inertia constant H of a 150-MVA, six-pole, 60-Hz synchronous machine is 4.2 MJ/MVA. Determine the value of WR^2 in lb · ft^2.

Ans. 1,893,939 lb · ft^2

10.21 The generator of Problem 10.20 is running at synchronous speed in the steady state. (*a*) What kinetic energy is stored in the rotor? (*b*) If the accelerating power due to a transient change is 28 MW, calculate the rotor acceleration.

Ans. (*a*) 630 MJ; (*b*) 480°/s^2

10.22 A 300-MVA, 1200-rpm synchronous machine has $WR^2 = 3.6 \times 10^6$ lb · ft^2. Calculate H for the machine (*a*) on its own base and (*b*) on a 100-MVA base.

Ans. (*a*) 3.99 MJ/MVA; (*b*) 11.97 MJ/MVA

10.23 A 100-MVA generator has $H = 4.2$ MJ/MVA, and 250-MVA machine, operating in parallel with the first, has $H = 3.6$ MJ/MVA. Calculate the equivalent inertia constant H for the two machines on a 50-MVA base.

Ans. 26.4 MJ/MVA

10.24 The moment of inertia of a 50-MVA, six-pole, 60-Hz generator is 20×10^3 kg · m^2. Determine H and M for the machine.

Ans. 3.15 MJ/MVA; 0.0146 MJ · s/degree

10.25 A synchronous motor develops 30 percent of its rated power for a certain load. The load on the motor is suddenly increased by 150 percent of the original value. Neglecting all losses, calculate the maximum power angle on the swing curve.

Ans. 40°

10.26 A 100-MVA synchronous generator supplies 62.5 MVA of power at 0.8 lagging power factor. The reactance between the load and the generator is normally 1.0 pu, but it increases to 3.0 pu because of a sudden three-phase short circuit. The fault is subsequently cleared and the generator then supplies 43.75 MVA at 0.8 lagging power factor. Determine the critical clearing angle.

Ans. 68.58°

10.27 A synchronous generator supplies its rated power to an infinite bus at a voltage of 1.0 pu. The reactance between the generator and the line, normally 0.825 pu, increases to 0.95 pu because of a fault. Find the critical clearing angle.

Ans. 58.73°

10.28 For the generator of Problem 10.15, determine the rotor speed in revolutions per minute at the end of ten cycles.

Ans. 1203.75 rev/min

10.29 A motor delivers 0.25 pu of its rated power while operating from an infinite bus. If the load on the motor is suddenly doubled, determine δ_m based on the equal-area criterion. Neglect all losses.

Ans. 45°

10.30 The inertia constant M of a synchronous machine is 4.45×10^{-4} pu. The machine operates at a steady-state power angle of 24.7°. Because of a fault, the power angle changes to a value given by the swing equation $\ddot{\delta} = 0.314$ pu. Using the step-by-step algorithm, plot the swing curve and use it to determine the maximum value of the power angle.

Ans. 67°

10.31 The *ABCD* constants for the nominal-Π circuit representation of a transmission line are $A = D = 0.9\underline{/0.3°}$, $B = 82.5\underline{/76°}\,\Omega$, and $C = 0.0005\underline{/90°}\,S$. What is the maximum power that can be transmitted over the line without making the system unstable if $|V_S| = |V_R| = 110$ kV?

Ans. 114.09 MW

10.32 Sketch the power-angle diagram for the line of Problem 10.31 when that line is represented by (*a*) an approximate series circuit and (*b*) a series reactance only. Determine the maximum power transmitted in each case.

Ans. (*a*) 111.18 MW; (*b*) 151.16 MW

10.33 The per-unit reactances for a given system are shown in Fig. 10-10. Unit power is being delivered to the receiving-end bus of the system at unity power factor and unit voltage. A three-phase short circuit occurs at F, the receiving end of one of the lines. Find the critical clearing angle.

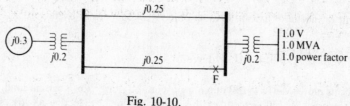

Fig. 10-10.

Ans. 59°

10.34 A 50-MVA, 33-kV, three-phase, four-pole, 60-Hz synchronous generator delivers 40 MW of power to an infinite bus through a total reactance of 0.55 pu. Because of a sudden fault, the reactance of the transmission line changes to 0.5 pu. The inertia constant of the machine 4.806 MJ/MVA. Sketch the swing curve during the fault, assuming that the voltage at the infinite bus is 1.0 pu and that behind transient reactance is 1.05 pu. The transient reactance of the machine is 0.4 pu.

Ans. Fig. 10-11

10.35 In a plant, two synchronous machines swing together. The inertia constants of the machines are H_1 and H_2, their MVA ratings are S_1 and S_2, the per-unit mechanical power inputs to the two units are P_{m1} and P_{m2}, and P_{e1} and P_{e2} are respectively the electrical power developed by the machines. Obtain an equivalent swing equation for the two-machine system in terms of inertia constants referred to a common base, the per-unit synchronous frequency ω_s in radians per second, the per-unit electrical frequency in radians per second, and the given values of per-unit power.

Ans. $\dfrac{2}{\omega_s}(H_1 + H_2)\omega_{\text{pu}}(t)\ddot{\delta} = P_{m1} + P_{m2} - (P_{e1} + P_{e2})$

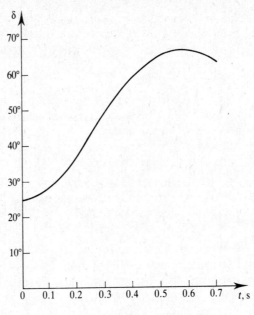

Fig. 10-11.

10.36 During a fault lasting 0.05 s, the swing equation for a 60-Hz maching was, for per-unit values,

$$\ddot{\delta} = \frac{5\pi}{3} \qquad 0 \leq t \leq 0.05 \text{ s}$$

The initial power angle was 0.418 rad. When the fault was cleared, the developed electrical power became $2.46 \sin \delta$ per unit. Determine (a) the maximum power angle and (b) whether or not the machine remained stable.

Ans.   (a) 156°; (b) remained stable

10.37 Calculate the critical clearing time in cycles for the machine of Problem 10.37.

Ans.   11.5 cycles

10.38 Rework Problems 10.36 and 10.37 using a numerical method.

Chapter 11

Power System Protection

We have seen in earlier chapters that a fault in a power system can lead to abnormal currents and voltages. For example, during a three-phase short circuit, the currents may become excessively large and the voltages may go to zero. The system must be protected against such occurrences, and steps must be taken to remove a fault as quickly as possible. In this chapter we examine some of the means for doing so.*

11.1 COMPONENTS OF A PROTECTION SYSTEM

Three types of components generally constitute a power system protection system: circuit breakers, transducers, and relays. In essence, when a fault occurs on the system, a voltage or current signal is transmitted to a relay by a transducer. The relay, in turn, operates a circuit breaker, and thereby the fault is cleared. The fault gives rise to abnormal voltages and currents, which may be in the range of kilovolts and kiloamperes. The transducer reduces them to much lower levels before transmitting the signal to the relay. The entire sequence of sensing and clearing the fault must be fast and reliable.

Figure 11-1 shows a one-line diagram of a portion of a power system with the components of its protection system in place.

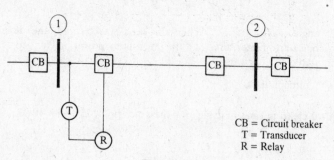

CB = Circuit breaker
T = Transducer
R = Relay

Fig. 11-1.

For reliability, the concept of *zones of protection* is implemented in protection systems. Figure 11-2 shows overlapping zones of protection, indicated by closed dashed lines, for a typical power system. Each zone contains two circuit breakers and one or more components of the power system. When a fault occurs within a zone, the protection system for that zone acts to isolate the zone from the rest of the system. The overlapping of zones ensures that no portion of the power system is left unprotected. However, the regions of overlap must be made as small as possible.

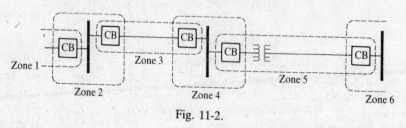

Fig. 11-2.

---

* Follows W. D. Stevenson, Jr., *Elements of Power System Analysis,* 4th ed., McGraw-Hill, 1982.

11.2 TRANSDUCERS AND RELAYS

As was mentioned earlier, transducers are used to reduce abnormal current and voltage levels and transmit input signals to the relays of a protection system. These transducers take the form of current and voltage (or potential) transformers, also known as *instrument* transformers. In contrast to power transformers, the power ratings of instrument transformers are rather low, perhaps 25 to 500 VA, depending on the load or *burden* on the transformer.

A *current transformer* (CT) is symbolically represented as in Fig. 11-3. The primary generally consists of the transmission line (*ab* in Fig. 11-3); the secondary winding consists of a multturn coil. The dots in the symbol imply that the secondary current leaving terminal *a'* is ideally in phase with the primary current entering terminal *a*. Nonideal instrument transformers have phase-angle and ratio errors, as shown in Fig. 11-4. Standard CT transformation ratios range from 50:5 to 1200:5.

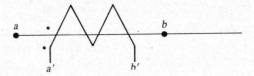

Fig. 11-3.

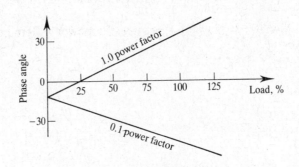

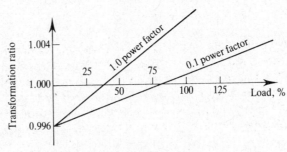

Fig. 11-4.

Voltage transformers (VTs) for application at or below 12 kV (primary voltage) generally have a 67-V secondary winding. For higher-voltage applications, a configuration of the type shown in Fig. 11-5 is used. In such a *coupling-capacitor voltage transformer* (CVT), with the appropriate *L* and *C* values (tuned for resonance), the phase-angle error is eliminated. Also, C_1 and C_2 are chosen so that only a few kilovolts appear across C_2 when *A* is at the (infinite) bus voltage, and the tapped voltage is reduced to the relay operating voltage.

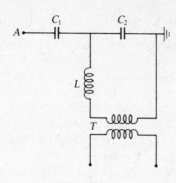

Fig. 11-5.

11.3 RELAY TYPES

The majority of the relays used in protection systems are of the following five types: magnitude relays, directional relays, ratio relays, differential relays, and pilot relays.

Magnitude relays, also known as *overcurrent relays,* respond to current inputs. They operate to trip a circuit breaker when the fault current exceeds a predetermined value. The current (on the secondary side of a CT) required to actuate the relay is known as the *pickup current* $|I_p|$. If $|I_F|$ is the fault current referred to the secondary, then the relay operates according to the following constraints:

$$\text{Trip for} \quad |I_F| > |I_p|$$
$$\text{Block for} \quad |I_F| < |I_p| \tag{11.1}$$

These constraints are shown graphically in Fig. 11-6. There is, however, another constraint—the relay operating time T, which is a function of I_F and I_p. That is, the time required for the relay to operate once $|I_F|$ exceeds $|I_p|$ may be written as the function

$$T = g(|I_F| - |I_p|) \tag{11.2}$$

and represented by a circle such as T_1 or T_2 in Fig. 11-6.

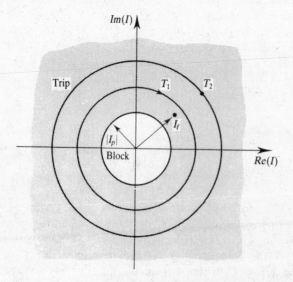

Fig. 11-6.

The time characteristics of overcurrent relays are more generally represented in the form of curves like those in Fig. 11-7. The pickup current is adjusted by choosing the proper primary tap setting. (We demonstrate the utility of these curves in Problem 11.2.)

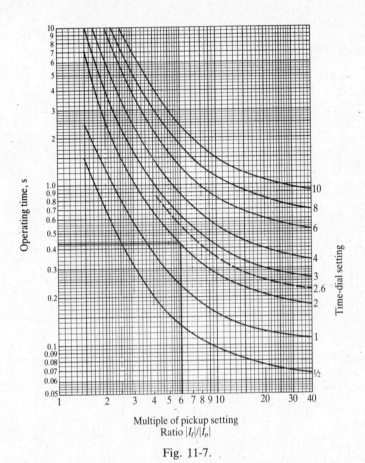

Multiple of pickup setting
Ratio $|I_f|/|I_p|$

Fig. 11-7.

A *directional relay* responds to faults either to the left or to the right of its location. Its operation depends upon the direction (lead or lag) of the fault current with respect to a reference voltage. If the reference voltage is V_{ref}, faults producing lagging currents in the shaded region of the phasor diagram of Fig. 11-8 will cause the relay to trip (and for all other faults it will block). The reference voltage is known as the *polarizing voltage*. The constraints on the operation of a directional relay are also given by

$$\text{Trip for} \qquad \theta_{\min} > \theta_{\text{op}} > \theta_{\max}$$
$$\text{Block for} \qquad \theta_{\min} < \theta_{\text{op}} < \theta_{\max}$$

(11.3)

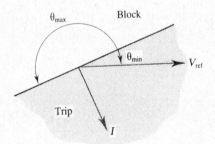

Fig. 11-8.

where θ_{op} is the phase angle of the operating quantity (I in Fig. 11-8), relative to that of the reference quantity (V_{ref} in the figure), and θ_{min} and θ_{max} define the boundaries of the operating range.

A *ratio relay* responds to faults only within a certain distance of its location. Suppose the impedance of a length of transmission line equal to that distance is $|Z_r|$, and denote the ratio of the voltage to the current at the location of the relay by Z (see Fig. 11-9). A relay with the operating constraints

$$\text{Trip for} \quad |Z| < |Z_r|$$
$$\text{Block for} \quad |Z| > |Z_r| \tag{11.4}$$

is called an *impedance* or *distance* relay and, in the complex impedance plane, has the operating characteristic shown in Fig. 11-9(*b*). Note that this relay is bidirectional. On the other hand, by offsetting the circle of Fig. 11-9(*b*) by Z', we obtain the relay constraints

$$\text{Trip for} \quad |Z - Z'| < |Z_r|$$
$$\text{Block for} \quad |Z - Z'| > |Z_r| \tag{11.5}$$

By selecting $|Z'|$ to be equal to $|Z_r|$, the relay characteristic can be made to pass through the origin, as illustrated in Fig. 11-9(*c*). Such a relay is obviously directional and is called a *mho* relay.

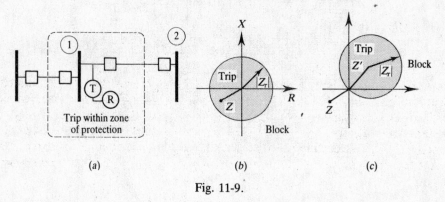

(*a*)   (*b*)   (*c*)

Fig. 11-9.

The operation of a *differential relay* may be understood by referring to Fig. 11-10. Under normal conditions we have $I_1 - I_2 = 0$. Under fault conditions, $I_1 - I_2 = I_F$, where I_F is the fault current referred to the secondary of the CTs. If a current $|I_p| < |I_F|$ is chosen to cause relay operation, then the relay's operating constraints are

$$\text{Trip for} \quad |I_1 - I_2| > |I_p|$$
$$\text{Block for} \quad |I_1 - I_2| < |I_p| \tag{11.6}$$

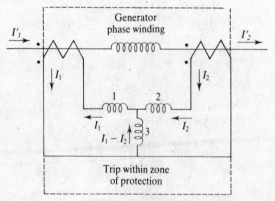

Fig. 11-10.

Note that the zone of protection of a differential relay is small; that is, the boundary points of the zone are closed to each other.

A *pilot relay* provides a means of transmitting fault signals from a remote zone boundary to relays at the terminals of a long transmission line.

11.4 PROTECTION OF LINES, TRANSFORMERS, AND GENERATORS

A radial transmission line like that of Fig. 11-11 can be protected with time-overcurrent relays. These relays can be set to provide primary protection for one line and remote backup protection for a neighboring line. For instance, the relay at bus 1 will protect the line from bus 1 to bus 2 and act as a backup for the line between buses 2 and 3. It must, however, be adjusted to provide an adequate time delay, such that the relay at bus 2 operates first for a fault on line 2.

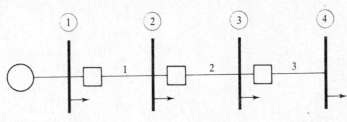

Fig. 11-11.

To protect lines fed from both ends (Fig. 11-12) or loop systems (Fig. 11-13), directional relays with coordinated time settings are used. (In the figures the arrows show the direction protected by each relay.) The relays associated with circuit breakers 1, 3, and 5 must be coordinated, as must the relays associated with breakers 2, 4, and 6. Overcurrent relays are used and are made directional by adding a directional relay at each location, and then arranging the outputs of the directional relay and the overcurrent relay so that their breakers will not operate unless both relays provide a trip signal.

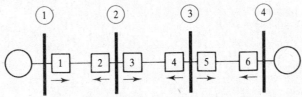

Fig. 11-12.

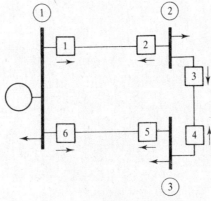

Fig. 11-13.

Transmission lines belonging to a complex interconnected system are protected by *impedance relays*, which respond to the impedance between their own location and the location of a fault.

Transformers and generators are protected against certain types of faults by differential relays. Figures 11-14 and 11-10, respectively, show arrangements of differential relays to protect against faults on a transformer and a generator.

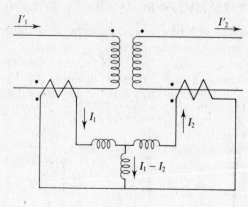

Fig. 11-14.

Solved Problems

11.1 A three-phase, delta-wye-connected, 30-MVA, 33/11-kV transformer is protected by a differential relay. Calculate the relay current setting for faults drawing up to 200 percent of the rated current. The CT current ratio on the primary side is 500:5, and that on the secondary side is 2000:5.

The primary line current is

$$I_p = \frac{30 \times 10^6}{\sqrt{3} \times 33 \times 10^3} = 524.88 \text{ A}$$

and the secondary line current is

$$I_s = 3I_p = 1574.64 \text{ A}$$

The CT current on the primary side is

$$I_1 = 524.88\left(\frac{5}{500}\right) = 5.249 \text{ A}$$

and that on the secondary side is

$$I_2 = 1574.64\left(\frac{5}{2000}\right)\sqrt{3} = 6.818 \text{ A}$$

The relay current at 200 percent of the rated current is then

$$2(I_2 - I_1) = 2(6.818 - 5.249) = 3.3138 \text{ A}$$

11.2 A portion of a radial system is shown in Fig. 11-15. For faults at bus 3, the maximum and minimum fault currents are 200 A and 165 A, respectively, and for fault at bus 2 the fault currents range between 300 A and 238 A. Using overcurrent relays having the characteristics shown in Fig. 11-7, with available tap settings at 3.0, 4.0, 5.0, 6.0, and 7.0 A, select the CT ratios, relay tap settings, and relay time-dial settings for the protection of the subsystem of

Fig. 11-15. The breaker at each bus opens all three phases when tripped by either of the two associated relays.

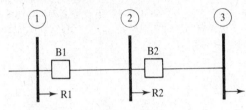

Fig. 11-15.

Settings for relay R2: We elect to provide a safety factor of 3, so R2 must operate when the line current is $\frac{1}{3}(165) = 55$ A. Choosing the closest standard CT ratio, which is 50:5, we obtain a relay current of $55(5/50) = 5.5$ A. Thus, we choose a tap setting of 5.0 A. We choose a time-dial setting of 1/2 for the fastest possible operation.

Settings for relay R1: There must be a relay R3 (not shown in the figure) that provides backup for R2. Relay R1 must then pick up reliably for the smallest current seen by R2. Consequently, we again use a CT ratio of 50:5 and a tap setting of 5A. As a first step in choosing a time-dial setting, we operate at least 0.3 s later than R2. We note that,the maximum fault current seen by R2 is 300 A. The relay current for both R1 and R2 is then $300(5/50) = 30$ A. For a relay tap setting of 5 A, the ratio of relay current to tap setting for both relays is $30/5 = 6.0$. In Fig. 11-7, we find that, for this ratio, the operating time for R2 (which has a time-dial setting of 1/2) is 0.135 s. Hence, if R2 fails, R1 must operate in $0.135 + 0.3 = 0.435$ s. Figure 11-7, shows that the required time-dial setting for R1 is 2.0.

11.3   A three-phase, delta-wye-connected, 15-MVA, 33/11-kV transformer is protected by CTs. Determine the CT ratios for differential protection such that the circulating current (through the transformer delta) does not exceed 5 A.

The line currents are

$$I_\Delta = \frac{15 \times 10^6}{\sqrt{3} \times 33 \times 10^3} = 262.44 \text{ A}$$

$$I_Y = \frac{15 \times 10^6}{\sqrt{3} \times 11 \times 10^3} = 787.30 \text{ A}$$

If the CTs on the high-voltage side are connected in wye, then the CT ratio on the high-voltage side is $787.30/5 = 157.46$.

Similarly, the CT ratio on the low-voltage side is $262.44(5/\sqrt{3}) = 757.6$.

Supplementary Problems

11.4   A three-phase, wye-delta-connected, 345/34.5-kV transformer has an emergency rating of 60 MVA. Determine the CT ratios and CT connections required for the protection of the transformer.

Ans.   1000:5, connected in wye on the low-voltage side; 200:5, connected in delta on the high-voltage side

11.5   What are the secondary currents in the CTs of Problem 11.4?

Ans.   5.0 A; 2.51 A

11.6   For faults at buses 2 and 3 of the radial system shown in Fig. 11-16, determine the CT ratio and relay setting for protective relays having tap settings of 5.0, 6.0, 7.0, 8.0, 10.0, and 12.0 A, assuming a safety factor of 1.3.

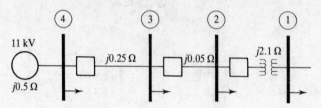

Fig. 11-16.

Ans. 1200:5; 12.0 A

11.7 The system shown in Fig. 11-17 is protected by relays having the characteristics shown in Fig. 11-7. The maximum and minimum fault currents are as follows:

	Fault at bus				
	1	2	3	4	5
Maximum fault current, A	3187.2	658.5	430.7	300.7	202.7
Minimum fault current, A	1380.0	472.6	328.6	237.9	165.1

Proceeding as in Problem 11-2, with available relay tap settings of 3.0, 4.0, 5.0, 6.0, and 7.0, determine the CT ratios, the relay tap settings, and the relay time-dial settings for the protection of the system.

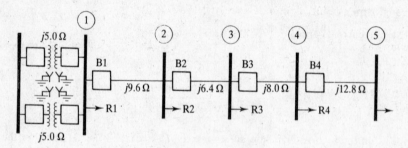

Fig. 11-17.

Ans.

	R1	R2	R3	R4
CT Ratio	100:5	100:5	50:5	50:5
Pickup setting, A	5	4	5	5
Time-dial setting	2.9	2.6	2.0	1/2

INDEX

173